Studies in Big Data

Volume 72

The series "Studies in Big Data" (SBD) publishes new developments and advances in the various areas of Big Data- quickly and with a high quality. The intent is to cover the theory, research, development, and applications of Big Data, as embedded in the fields of engineering, computer science, physics, economics and life sciences. The books of the series refer to the analysis and understanding of large, complex, and/or distributed data sets generated from recent digital sources coming from sensors or other physical instruments as well as simulations, crowd sourcing, social networks or other internet transactions, such as emails or video click streams and other. The series contains monographs, lecture notes and edited volumes in Big Data spanning the areas of computational intelligence including neural networks, evolutionary computation, soft computing, fuzzy systems, as well as artificial intelligence, data mining, modern statistics and Operations research, as well as self-organizing systems. Of particular value to both the contributors and the readership are the short publication timeframe and the world-wide distribution, which enable both wide and rapid dissemination of research output.

** Indexing: The books of this series are submitted to ISI Web of Science, DBLP, Ulrichs, MathSciNet, Current Mathematical Publications, Mathematical Reviews, Zentralblatt Math: MetaPress and Springerlink.

More information about this series at http://www.springer.com/series/11970

Pradeep Kumar Garg · Rahul Dev Garg ·
Gaurav Shukla · Hari Shanker Srivastava

Digital Mapping of Soil Landscape Parameters

Geospatial Analyses using Machine Learning and Geomatics

Springer

Pradeep Kumar Garg
Geomatics Section, Civil Engineering Department
Indian Institute of Technology Roorkee
Roorkee, Uttarakhand, India

Rahul Dev Garg
Geomatics Section, Civil Engineering Department
Indian Institute of Technology Roorkee
Roorkee, Uttarakhand, India

Gaurav Shukla
Surveying and Geomatics Section, Civil Engineering Department
Maharishi Markandeshwar University
Ambala, Haryana, India

Hari Shanker Srivastava
Indian Institute of Remote Sensing (IIRS)
Dehradun, Uttarakhand, India

ISSN 2197-6503 ISSN 2197-6511 (electronic)
Studies in Big Data
ISBN 978-981-15-3237-5 ISBN 978-981-15-3238-2 (eBook)
https://doi.org/10.1007/978-981-15-3238-2

This Springer imprint is published by the registered company Springer Nature Singapore Pte Ltd.
The registered company address is: 152 Beach Road, #21-01/04 Gateway East, Singapore 189721, Singapore

About This Book

This book begins with the fundamental concepts behind the mapping of soil-landscape parameters in geospatial domain. As readers go through the chapters of this book, they will learn how machine learning and geomatics can be applied for more efficient mapping, understanding, and management of 'soil'.

The judicious utilization of the piece of land is the biggest and most important challenging issue. This is especially true in light of the rapid urbanization worldwide, which requires continuous monitoring of resource consumption. This book provides a clear idea of how machine learning can be used to explore a remote sensing data to monitor the key parameters, below the surface, at the surface, and above the surface. Not only will readers gain insight into the approaches, they will also learn challenges and issues associated with the digital mapping of these parameters. This book also provides understanding, into the selection of data set to represent soil–landscape relationships. Readers will also be able to understand the complex and interconnected linkages between soil-landscape parameters under a range of soil and climatic conditions. The readers gain insight into utilization of a network of satellite-based Earth observations to provide a solution towards smart farming and smart land management.

The first two chapters prepare the readers to comprehend the basic concepts, approaches, and issues/challenges associated with digital mapping of soil parameters. All other chapters develop reader's ability to handle the complexity involved in the digital mapping of three key parameters, i.e. below the surface: soil moisture, at the surface: soil class, and above the surface: crop cover.

Our potential readers will be researchers and geo-scientist, exploring digital mapping techniques for resource management, environmental planning, and sustainable development, as well as the college and university students studying geospatial technologies. It is also hoped that book contents will also attract and inspire the individuals now in the industry who requires framework in the field related to smart land planning, crop management, precision agriculture (satellite farming), and sustainable-city development in geospatial domain.

Contents and Coverage

The book comprises 6 chapters. Chapter 1: "Concept of Digital Mapping"—covers fundamental of digital mapping, key soil-landscape parameters, different approaches, and important challenging issue in digital mapping of soil-landscape parameters. Chapter 2: "Different Approaches on Digital Mapping of Soil-Landscape Parameters"—provides an in-depth explanation of various approaches for the mapping of soil classes, crop cover, and soil moisture using different machine learning algorithms. It also covers the selected method, structure, and assessment systems. Chapter 3: "Selection of Suitable Variables and Their Development"—covers the suitable variables and their development related to soil properties and parent material, climate, organisms/vegetation, relief/topography, relative position, soil moisture, and surface characteristics. Satellite data and its pre-processing are also presented for development of suitable variables. Chapter 4: "Digital Soil Mapping: Implementation and Assessment"—is devoted to approach and assessment of digital soil mapping. It discusses the various approaches that have been, or could be, used for fitting relationships between soil properties or classes and soil-forming factors. It also discusses the data layers that have been, or could be, used to describe the soil-forming factors and their importance in digital soil mapping. Chapter 5: "Prediction Models for Crop Mapping"—describes the framework for implementation of models for crop-type prediction using climate variables in association with texture, vegetation, soil moisture, phenology, topography, parent material, and soil, followed by comparative performance analysis of different ensemble models. It also discusses the potential of different variables for crop-type classification. Chapter 6: "Spatial Soil Moisture Prediction Model Over an Agricultural Land"—discusses the framework for surface soil moisture prediction model. It also discusses the model evaluation based on the goodness-of-fit test and probability of non-exceedance on field and SMAP (Soil Moisture Active and Passive) data.

A Bibliography at the end of each chapter is for the benefits of the students.

Contents

About the Authors

Professor Pradeep Kumar Garg is faculty in Civil Engineering Department of Indian Institute of Technology, Roorkee. He has also served as vice chancellor, Uttarakhand Technical University, Dehradun. He completed his B.Tech. in 1980 and his M.Tech. in 1982, both from the University of Roorkee, India (now IIT Roorkee). He did a Ph.D. from University of Bristol, UK, and postdoctoral research work at the University of Reading, UK. He joined the Department of Civil Engineering at IIT Roorkee in 1982. Dr. Garg has published about 93 research papers, guided 7 Ph.D. thesis, 47 M.Tech. thesis, authored one textbook on Remote Sensing, organized 22 training programmes, and prepared two educational films under UGC programme. He has completed 13 research projects and 25 consultancy projects. He is Fellow member of 5 Technical Societies and life member of 15 Technical Societies. His main areas of research interest are remote sensing and GIS applications.

Professor Rahul Dev Garg is faculty in Civil Engineering Department of Indian Institute of Technology, Roorkee. He graduated with a bachelor's in technology in Civil Engineering in 1989, master's in technology in 1993, and a Ph.D. in 2004 from IIT Roorkee. He has also served as a scientist from 1993 to 2007 in Indian Institute of Remote Sensing (IIRS), Dehradun. Dr. Garg has published about 110 research papers, guided 8 Ph.D. thesis, 47 M.Tech. thesis, authored one textbook on Remote Sensing, organized 22 training programmes, and prepared two educational films under UGC programme. He has completed 11 research projects and 24 consultancy projects. He is Fellow member of 3 Technical Societies and life member of 10 Technical Societies. His main areas of interest are land surveying, remote sensing, GIS, GPS, digital image processing, SAR interferometry, and GPR.

Dr. Gaurav Shukla is currently working as a faculty member in surveying and geomatics section, Civil Engineering Department, Maharishi Markandeshwar (Deemed to be University) University, Mullana, Haryana, India. He is also a nodal coordinator of MHRD initiative, virtual lab programme. He is postgraduated in geomatics from the Indian Institute of Technology (ISM), Dhanbad, in 2011 and completed his Ph.D. from Indian Institute of Technology, Roorkee, India, in 2018. Dr. Shukla has published 7 journal papers and organized 2 training programmes. His main areas of research interest include nonparametric approaches to retrieval of Earth's parameters, GNSS reflectometry, remote sensing, and GIS applications.

Dr. Hari Shanker Srivastava is a senior scientist, faculty and Group Head, Programme, Planning and Evaluation Group (PPEG) of Indian Institute of Remote Sensing (IIRS), ISRO, Dehradun, India. He is gold medallist in M.Sc. Physics and Ph.D. in Physics on Synthetic Aperture Radar (SAR). He is Ph.D., M.Tech., M.Sc. and B.Tech. guide at IIRS, IIT-Dhanbad, IIT-Roorkee, IIT-BHU, BITS, Pilani etc. His research interest is in the field of Microwave Remote Sensing focusing on multi-parametric SAR, passive microwave radiometers, Interferometry, Polarimetry and Polarimetric Interferometry for soil moisture, agriculture, wetland, forestry, human settlement, radiometric calibration of active sensors and validation of soil moisture derived from passive sensors. With more than 150 research publications to his credit, he is reviewer to a number of reputed journals published by IEEE, Elsevier, Taylor & Francis, Springer, SPIE etc.

Abbreviations

ALOS	Advanced Land Observing Satellite
ANN	Airborne Visible/Infrared Imaging Spectrometer
ASTER	Advanced Spaceborne Thermal Emission and Reflection Radiometer
AVHRR	Advanced Very High Resolution Radiometer
CART	Classification and Regression Tree
CDT	CART Decision Trees
CEA	CART Ensembles ARCing
CEB	CART Ensemble Bagger
CKC	Cohen's Kappa coefficient
CTI	Compound Topographic Index
DEM	Digital Elevation Model
DSM	Digital Soil Mapping
DTM	Digital Terrain Model
EM	Electromagnetic
GIS	Geographical Information System
GPS	Global Positioning System
IRS	Indian Remote Sensing
ISLE	Importance Sampled Learning Ensemble
JCC	Jaccard's Coefficient of Community
jNSM	Java Newhall Simulation Model
LISS	Linear Imaging Self Scanning Sensor
LR	Linear Regression
MDA	Mean Decrease in Accuracy
MDG	Mean Decrease in Gini
MLR	Multinomial Logistic Regression
MODIS	Moderate Resolution Imaging Spectroradiometer
NDVI	Normalized Difference Vegetation Index
NPV	Negative Predictive Value
OLI	Operational Land Imager
PPV	Positive Predictive Value

RADAR	Radio Detection and Ranging
RF	Random Forest
RH	Right Circular Transmit and Horizontal Receive
RISAT	Radar Imaging Satellite
RS	Remote Sensing
RV	Right Circular Transmit and Vertical Receive
SAR	Synthetic Aperture Radar
SMAP	Soil Moisture Active and Passive
SMPM	Soil Moisture Prediction Model
SMR	Soil Moisture Regime
SRTM	Shuttle Radar Topography Mission
SVM	Support Vector Machine
TNGBM	TreeNet Gradient Boosting Machine
TNR	True Negative Rate
TPR	True Positive Rate
USGS	United States Geological Survey
UTM	Universal Transverse Mercator projection
VH	Vertical Transmit and Horizontal Receive
VV	Vertical Transmit and Vertical Receive

List of Figures

List of Tables

Chapter 1
Concept of Digital Mapping

1.1 Key Soil-Landscape Parameters

Managing the land is the biggest and most important challenging issue being faced in today's rapidly evolving world. It is necessary to plan smartly with Earth's limited land to grow enough food, produce enough energy, and still preserve some of the species (both plant and animal) on the planet. The United Nation's latest estimates state that there are 7.6 billion people jostling for space on Earth at present and that number will rise to 9.8 billion by 2050. By the end of the century, there could be 11.2 billion people on our planet (Gray 2017). With 83 million more people appearing on the planet every year, rising populations are placing increasing pressure on the land. Each of those people will need somewhere to live, a place to work, fertile land to provide them with food. But for all of them, there can be no doubt the impact they will all have on the land and those impacts will be one of the grand challenges faced by humanity in the coming century.

Therefore, there is a clear need for more efficient mapping, understanding, and management of 'soil' in meeting the increasing demand for food and other basic requirements of an ever-increasing population of the world. Mapping the spatial distribution of soil-landscape parameters is important for informing soil use and management decisions. The key parameters, *below the surface*, *at the surface,* and *above the surface* are dynamic under a range of soil and climatic conditions (Holah et al. 2005; Giasson et al. 2006; Schrott and Sass 2008; Zhang et al. 2011). These parameters vary in terms of presence and composition from landscape to landscape, which help to define the characteristics of any particular soil-landscape. All possible necessary information on the land resources should be collected to enable the land use planner to prepare a scientific land use plan such that the piece of land should be used for only that purpose for which it is best suitable. In this way, maximum possible return from every piece of land can be obtained without any damage or decrease in its productivity. The people should also be properly educated to avoid misuse of the land. Unless the available productive land, which is the main source

P. K. Garg et al., *Digital Mapping of Soil Landscape Parameters*, Studies in Big Data 72,
https://doi.org/10.1007/978-981-15-3238-2_1

of human sustenance is protected and used in a scientific way to give better and increased returns, there is no hope of human survival in the future.

Three key parameters, i.e. *below the surface*: soil moisture, *at the surface*: soil class, and *above the surface*: crop cover are selected for prediction modelling using space-based technologies. These parameters are closely interrelated. The vegetation (either natural vegetation or agricultural crops) above surface produces a large amount of organic matter and accelerates biological activity (Yemefack 2005; Gomi et al. 2008; Schoonover and Crim 2015). Falling leaves decompose to form the humus, which accelerates the weathering of primary minerals of the soil to form more clay (Little et al. 1998; Körschens 2002; Gomi et al. 2008; Kravchenko 2008). The vegetation and leaf litter cover also protect the soil surface from direct rain, reduces runoff on steep slope thereby allowing more infiltration. However, in the hot dry season, the bare soil has diminished biological activity, production of organic matter is suspended, mineralization of organic matter is intense, and degradation of soil structure to prisms or crusts results. Such soils with degraded topsoil layer are more erodible because of low porosity and reduced moisture-holding capacity (Körschens 2002; Gomi et al. 2008; Vishwanath et al. 2011). Therefore, in this study, these three key parameters, i.e. *below the surface*: soil moisture, *at the surface*: soil class, and *above the surface*: crop cover are considered as each parameter plays an important role in the formation of soil-landscape of the region. Inter-correlation is the main basis for the selection of these parameters for the study.

Mapping of these parameters is also essential for agriculture purposes because different crops need different types of soils, different types, and amounts of nutrients and water (Mohler and Johnson 2009). By selecting the right crop for the given soil conditions, moisture conditions, and climate, one can optimize yields and save water requirements for irrigation (Mohler and Johnson 2009). Furthermore, it is widely recognized that soil survey and soil map are essential prerequisite for the planned use of land resources. Soil survey provides relevant information on the types of soils and their distribution. It is essential to have soil maps for different types of planning and uses, such as forest, farming, and grazing, engineering. An accurate and consistently prepared soil map will be very much helpful for planners and general users.

The broad general objectives of the soil surveys are both of fundamental and objective nature (Carre et al. 2007; Taghizadeh-Mehrjardi et al. 2014; Vaysse and Lagacherie 2015). Fundamentally, soil survey helps in expanding the knowledge and understanding of soils regarding their genesis, development, classification, and nomenclature. The applied part of soil survey includes interpretation of soil data for use of agriculture, forestry, pasture development, recreation, and others (Carre et al. 2007; Vaysse and Lagacherie 2015). It gives the information needed for planning land use development under irrigation and the management of irrigated lands. Further, it correlates with characteristics of soils of known behaviour and predicts their adaptability to various uses and also their behaviour and productivity under defined sets of management practices (Carre et al. 2007). So, it forms the very basis for planned land use. Similarly, the crop classification maps are useful for estimating the yield amount or the acreage of tillable land and managing the agricultural fields (Mohler and Johnson 2009).

Although only a tiny portion of total water stored in the soil [0.0012%, (Seneviratne et al. 2010; Cui et al. 2016)], surface soil moisture is one of the crucial variables in hydrological process, which influences the exchange of water and energy fluxes at the land surface/atmosphere interface (Wang and Qu 2009). Surface soil moisture is the water present in the upper 10 cm of soil, whereas root zone soil moisture is the water that is available to plants, which is generally considered to be in the upper 200 cm of soil. The knowledge of its spatial and temporal variation has importance with respect to study of the different processes on Earth's surface (Hallikainen et al. 1985; Dong 2011; Oltra-Carrió et al. 2015; Cui et al. 2016). The estimation of soil moisture by means of remotely sensed observations is very attractive to domains, such as hydrology, agronomy, or meteorology.

1.2 Retrieval Approaches

In summary, mapping of selected parameters is a prerequisite for land planning and management for planners as well as for common persons. The retrieval method is the core of a mapping system. It assumes that the addressed retrieval problem can be expressed in terms of mapping between a set of values of features extracted from the signals acquired using remote sensors and the desired continuous variable that is related to the target characteristics. From an analytic viewpoint, this concept can be expressed as (Ali et al. 2015):

$$Y = F(x) + e \tag{1.1}$$

where 'F' denotes the desired and unknown mapping and 'e' is a random variable taking into account all of the random noise contributions affecting the retrieval problem. From the methodological perspective, the retrieval of 'Y' corresponds to the problem of determining a mapping $F*$ as close as possible to the true mapping 'F'.

Traditional methods (e.g. field surveys, literature reviews, map interpretation, and collateral and ancillary data analysis) are not effective because they are time-consuming, date lagged, and often too expensive. In classical retrieval methods for soil-landscape parameters, the following three approaches have been addressed: (i) the derivation of empirical data-driven relationships, (ii) the inversion of physical models, and (iii) semi-empirical method to simplify the complexity of the physical model. The first approach relies on the availability of a set of reference samples, i.e. couples of in situ measurements of the desired target variable associated with the corresponding measurements of the remote sensor. These samples are exploited for deriving an empirical mapping, e.g. by means of statistical regression techniques in combination with parametric functions (linear, logarithmic, or polynomial). Then, the identified relationship is extended to the entire satellite image. An important issue in the empirical method is that the empirical relationships are typically site and sensor dependent since they are derived from samples collected under specific operational conditions. This limits the possibility to extend their use to different areas and different remote sensing systems since they remain valid only under the conditions in which reference samples have been collected.

The second approach demands the definition of the desired mapping function to analytic Electromagnetic (EM) models. Such models are based on a solid physical description of the mechanisms involving the interaction of EM radiation and the target object of interest. In the direct operational way, they simulate the response of a target object as a function of (i) the target characteristics (i.e. structural, chemical, and biophysical variables), and (ii) the signal characteristics (i.e. wavelength, incidence/reflection angle, etc.). Thus, in the inverse operational way, they can be used to represent the mapping between the measurements at the remote sensor and the variable of interest. A wide variety of analytic EM models have been proposed in the literature with different levels of complexity and generality.

A physical model for vegetation cover is based on the variations of vegetation reflectance with sun zenith angle, view zenith angle, and terrain slope angle. They either consider the vegetation canopy to comprise thin layers of leaves, suspended in the air like sediment particles in water forming a turbid medium (Kuusk 1995; Andrieu 1997) or the large shadows cast on the canopy surface (Li and Strahler 1985; 1992). Hybrid models combine the two approaches (Chen and Leblanc 1997). Of particular interest are the physically based models that through simplifying assumptions about the vegetation canopy, result in an explicit analytical formula for reflectance, like the empirical models, and thus potentially combine the advantages of the empirical and physical models.

The Universal Soil Loss Equation (USLE), and the Revised Universal Soil Loss Equation (RUSLE) are the empirical soil loss models. The USLE is the most widely known and used soil loss equation in the world (Ahmad et al. 2010; Aiello et al. 2015). The RUSLE can be described as a software version of a greatly improved USLE. Many studies integrated empirical model with GIS (Geographical Information System) and remote sensing imageries to find soil loss or erosion (Prasannakumar et al. 2012; Aiello et al. 2015; Ganasri and Ramesh 2016). GIS helped to integrate, analyse, and map potential soil erosion zones. Because of the potential capacity for systematic observations at various scales, remote sensing technology makes available possible data archives from the present time to over several decades back. Using these characteristics, enormous efforts have been made by researchers and application specialists to retrieve soil-landscape parameters from local scale to global scale by applying remote sensing imagery (Goel and Pal 2009; Ahmad et al. 2010; Coopersmith et al. 2016; Ganasri and Ramesh 2016).

Active microwave and passive optical remote sensing data have demonstrated their respective advantages in an inversion of surface soil moisture content. Optical remote sensing methods mainly consist of various Vegetation Index Models, Vegetation temperature Index Models, Thermal Inertia Models, and Ground Evaporation Models (Palacios-Orueta et al. 2005; Khanna et al. 2007; Giraldo et al. 2008; Zhang et al. 2008). These are empirical or semi-empirical models which use reflective spectral changes of the ground to simulate relationships among vegetation index, land surface temperature, soil thermal inertia, ground evaporation, and soil moisture content and hence estimate the surface soil moisture content. The microwave remote sensing methods are mainly physical models and include the Kirchhoff Model (KM), Small Perturbation Model (SPM), Integral Equations Model (IEM), and Michigan Microwave Canopy Scattering Model (MIMICS) (Dobson and Ulaby 1981; Ulaby et al. 1983; Dobson et al. 1985; Hallikainen et al. 1985; Du et al. 2000; Lu et al. 2009). The principle of these models is that soil moisture is linked to radiometric signatures by the dielectric constant. In addition, the radiometric signatures are influenced by surface roughness and vegetation cover effects. Currently, integration of optical and microwave remote sensing is attracting attention of researchers.

Physical models involve complicated scattering phenomena from probabilistic models of soil, vegetation, and terrain whereas empirical models capture relationships among measured variables to estimate geophysical characteristics. Even though, empirical models are data-driven and are not computationally intensive and complex; for both kinds of models, in situ data is required to calibrate and validate them.

With the advancements in machine learning, there are many techniques available nowadays to develop these models. Machine learning may be defined as the automated process to get the unknown relationship between input (predictor variable) and target parameters using a known example (training data) or intrinsic structures in data (Witten and Frank 2005). Machine learning uses two types of techniques: supervised learning, which trains a model on known input and output data so that it can predict future outputs, and unsupervised learning (clustering), which finds hidden patterns or intrinsic structures in input data (Breiman 2001; MathWorks 2017).

The main objective of supervised machine learning is to estimate the unknown relationship between input and target parameters using known examples. Then, the relationship thus derived can be used in predicting the unknown target values for other values of inputs. Supervised learning uses classification and regression techniques to develop predictive models based on target character. The targets can be nominal or numerical. If the targets are nominal, then the problem becomes a classification one while if the target is numerical, the problem is a regression one. Classification models classify input data into categories. Typical applications include: soil category of land, land use land cover, urban–nonurban area, etc. Regression techniques predict continuous responses and typical applications include soil moisture condition, drought monitoring, weather prediction, pollution level, etc.

Unsupervised learning finds hidden patterns or intrinsic structures in data. It is used to draw inferences from data sets consisting of input data without labelled responses. Clustering is the most common unsupervised learning technique (McBratney et al. 2003; Ali et al. 2015; MathWorks 2017). It is used for exploratory data analysis to find hidden patterns or groupings in data. Applications for clustering include gene sequence analysis, market research, and object recognition. Generally, researchers consider using machine learning when the problem is complex, and involve a large amount of data, a lot of variables, but no existing formula or equation. There are dozens of supervised and unsupervised machine learning algorithms, and each takes a different approach to learning. There is no best method or one-size-fits-all (Sutton 2004; Gorthi 2011; MathWorks 2017). Finding the right algorithm is partly just trial and error, and usually based on size and type of data, the insights user wants to get from the data, and how those insights will be used (Gorthi 2011).

Many algorithms can be used during development of a model in supervised learning, which are developed in different mathematical backgrounds, i.e. Support Vector Machines (SVM), Discriminant Analysis, Naive Bayes, Nearest Neighbor, Ensemble Methods, Decision Trees, Artificial Neural Networks (ANN), Linear Regression (LR), Random Forest (RF), etc., whereas K-Means, K-Medoids, Fuzzy C-Means, Hidden MarkovModel, and Gaussian Mixture, etc. are some prominent unsupervised clustering methods. Development of models using supervised learning involves the following procedure, i.e. (i) Identification of input parameters and collection of sample data for training and testing, (ii) Analyse the correlation between the input parameters. Less correlated inputs will give more reliable results, (iii) Selection of the algorithm and optimum parameter selection or model tune-up, (iv) Train the model using the training examples, and (v) Reliability and accuracy check by predicting the target values for the test data set. However, space-based environment retrieval methods have challenges associated with acquisition techniques and analytical algorithms. Complex and interconnected linkages between soil-landscape parameters under a range of soil and climatic conditions are a key open question for the scientific community.

1.3 Underlying Concepts and Challenges

Remote sensing of the environment is usually concerned with determination of the state of one (or several) physical, biological, or geographic parameters, such as soil temperature, soil moisture, the biomass (crop or forest), the quantity of a polluting gas in the atmosphere, the extent and the state of the global snow cover or land cover or soil categories, etc. (Zhang et al. 2011). When an EM wave interacts with matter some fraction of energy is transmitted, absorbed, or reflected. Changes in the chemical, physical, and structural characteristics of a matter (target) determine the variations of its EM response in terms of absorption, emission, transmission, and reflection (Ulaby and Batlivala 1976; Bell et al. 1986; Chanasyk and Naeth 1996). EM waves communicate this information to a remote sensor. If this sensor is able to receive and detect the information, i.e. if it is sensitive to the relevant part of EM waves, it can determine the EM properties of the target.

The major challenge of remote sensing is that the observations are not direct measurements of the parameters. A retrieval problem is one of findings the inverse of a forward function which describes the relation between the measurements and the atmospheric parameters of interest and therefore the data analysis involves an 'inverse' or 'retrieval' problem.

1.3.1 Challenges Associated with Acquisition Techniques

- *Selecting an observation frequency*: Selecting an observation frequency (such as visible, infrared, or microwave) is one the most critical factors for remote sensors. The observation frequency is closely related to how accurately you can extract, through various kinds of noise, variation of interaction of EM waves against a target object. If a study uses two or more different EM waves that interact differently with a target object, it allows more detailed characterization of the object (Ulaby and Batlivala 1976; Chanasyk and Naeth 1996).
- *The influence of external disturbing factors*: The remote sensing acquisition system is not ideal, but affected by disturbing factors, such as the noise and nonlinearity at the sensor level and the presence of the atmosphere. Particles and gases (water vapour, carbon dioxide, and ozone) in the atmosphere affect the incoming light and radiation and cause degradation (hazy or blurring appearance) of the remotely sensed images. Furthermore, the light from a target outside the field of view of the sensor may be scattered into the field of view of the sensor (Wang 1987). Even if these issues can be determined and corrected to some extent with the help of calibration and atmospheric correction procedures, they may still corrupt the signal measured at the sensor level and, thus, introduce further ambiguity and complexity in the retrieval process (Wang 1987; Shi et al. 1997; Hadjimitsis et al. 2004).
- *The image formation process at the sensor level*: In remote sensing, a digital number corresponding to any pixel, represents the average of several spectral classes within the area that it covers on the ground, each emitted or reflected by a different type of material. This behaviour is the origin of a mixed contribution at the sensor level. Even by increasing the spatial resolution, this mixing phenomenon cannot be completely cancelled, as mixed pixels are common along the edges of features. Moreover, a response corresponding to a pixel can also be affected by radiation components coming from the surroundings of the investigated area, thus, introduces complexity in the retrieval process (Wang 1987).

1.3.2 Challenges Associated with Analytical Algorithms

- One of the objectives of image analysis is to estimate the unknown relationship between input and target parameters using known examples. The relationship between remote sensing measurements and target variables is a very complex and nonlinear in nature. Because, geo-/biophysical variables may affect the EM properties of a target differently along with their range of variability, potentially leading to signal saturation and other nonlinear effects (Haboudane et al. 2004). Furthermore, EM radiation usually shows a non-uniform sensitivity to the different physical phenomena depending, for instance, on the wavelength of the signal or the acquisition geometry (Haboudane et al. 2004; Long et al. 2013; Sonobe et al. 2017).

- *The complexity of the retrieval problem*: The total EM response of a target is typically the result of multiple contributions, each one determined by a different structural, chemical, or physical characteristic (Wang 1987). Spectral response can be quite variable, even for the same target type, and can also vary with time (e.g. 'green-ness' of leaves) and location. Furthermore, the same EM responses can be associated with different geo-/biophysical variable configurations. Therefore, retrieval models have the issue of variable equifinality and parameter uncertainty, or parameter ambiguity. Problems of equifinality are common in retrieval models for use over mixed landscapes of natural and agricultural areas, for which few spatially distributed ground data are available (Whiting et al. 2006).
- It complicates the choice of model, makes the use of models for hypothesis testing difficult, and results in uncertain prediction for times or places that have not been used for calibration.
- Furthermore, analysis of temporal changes is more complex because it generally involves changes in both the target and imaging condition at the time of each acquisition (Chanasyk and Naeth 1996).
- *Suitable input variable selection*: The identification of important and redundant inputs is a critical issue that leads to reduced size of problem, faster training, and possibly more accurate results (Jackson and Schmugge 1991; Bui and Moran 2003; Jafari et al. 2013; Poggio et al. 2013).

Generally, within scientific community, retrieval problem is considered as an ill-posed problem, i.e. either there is not a unique mathematical solution or there is no solution at all. Because all real measurements include error there is non-uniqueness in the observations which leads to further non-uniqueness in the solution. The implications of this must be understood in order to make proper use of the data. The key parameters, below the surface, at the surface, and above the surface are dynamic under a range of soil and climatic conditions. Therefore, the identification of important and redundant inputs is a critical issue that leads to the reduced size of the problem, faster training, and possibly more accurate results. Potentiality of effective indicators is also not analysed to improve the understanding of the interactive feedback between each parameter and changes of soil and climatic conditions. Lack of effective framework for digital mapping and assessment of essential parameters in the spatial domain, in the Indian context. Digital soil mapping and space techniques can accelerate the provision of fine-scale information on soil fertility and conditions in India. It can also be used for monitoring and understanding soil nutrient change over time and for landscape planning. However, complex and interconnected linkages between soil-landscape parameters under a range of soil and climatic conditions are key open questions for the scientific community.

References

Ahmad, S., Kalra, A., & Stephen, H. (2010). Estimating soil moisture using remote sensing data: A machine learning approach. *Advances in Water Resources, 33*(1), 69–80.

Aiello, A., Adamo, M., & Canora, F. (2015). Remote sensing and GIS to assess soil erosion with RUSLE3D and USPED at river basin scale in southern Italy. *CATENA, 131*, 174–185.

Ali, I., Greifeneder, F., Stamenkovic, J., Neumann, M., & Notarnicola, C. (2015). Review of machine learning approaches for biomass and soil moisture retrievals from remote sensing data. *Remote Sensing, 7*(12), 16398–16421.

Andrieu, B. (1997). Evaluation of an improved version of SAIL model for simulating bidirectional reflectance of sugar beet canopies. *Remote Sensing of Environment, 60*(3), 247–257.

Bell, F. G., Cripps, J. C., & Culshaw, M. G. (1986). A review of the engineering behaviour of soils and rocks with respect to groundwater. *Geological Society, London, Engineering Geology Special Publications, 3*(1), 1–23.

Breiman, L. (2001). Random forest. *Machine Learning, 45*(1), 5–32.

Bui, E. N., & Moran, C. J. (2003). A strategy to fill gaps in soil survey over large spatial extents: An example from the Murray-Darling basin of Australia. *Geoderma, 111*(1–2), 21–44.

Carre, F., McBratney, A. B., Mayr, T., & Montanarella, L. (2007). Digital soil assessments: Beyond DSM. *Geoderma, 142*(1–2), 69–79.

Chanasyk, D. S., & Naeth, M. A. (1996). Field measurement of soil moisture using neutron probes. *Canadian Journal of Soil Science, 76*(3), 317–323.

Chen, J. M., & Leblanc, S. G. (1997). A four-scale bidirectional reflectance model based on canopy architecture. *IEEE Transactions on Geoscience and Remote Sensing, 35*(5), 1316–1337.

Coopersmith, E. J., Cosh, M. H., Bell, J. E., & Boyles, R. (2016). Using machine learning to produce near surface soil moisture estimates from deeper in situ records at U.S. Climate Reference Network (USCRN) locations: Analysis and applications to AMSR-E satellite validation. *Advances in Water Resources, 98*, 122–131.

Cui, Y., Long, D., Hong, Y., Zeng, C., Zhou, J., Han, Z., et al. (2016). Validation and reconstruction of FY-3B/MWRI soil moisture using an artificial neural network based on reconstructed MODIS optical products over the Tibetan Plateau. *Journal of Hydrology, 543*, 242–254.

Dobson, M. C., & Ulaby, F. T. (1981). Microwave backscatter dependence on surface roughness, soil moisture and soil texture: Part III—Soil tension. *IEEE Transactions on Geoscience and Remote Sensing, 19*(1), 51–61.

Dobson, M. C., Ulaby, F. T., Hallikainen, M. T., & El-Rayes, M. A. (1985). Microwave dielectric behavior of wet soil-part II: Dielectric mixing models. *IEEE Transactions on Geoscience and Remote Sensing, GE-23*(1), 35–46.

Dong, L. (2011). *Operational retrieval of surface soil moisture using synthetic aperture radar imagery in a semi-arid environment*. Ludwig Maximilian University of Munich.

Du, Y., Ulaby, F. T., & Dobson, M. C. (2000). Sensitivity to soil moisture by active and passive microwave sensors. *IEEE Transactions on Geoscience and Remote Sensing, 38*(1), 105–114.

Ganasri, B. P., & Ramesh, H. (2016). Assessment of soil erosion by RUSLE model using remote sensing and GIS—A case study of Nethravathi Basin. *Geoscience Frontiers, 7*(6), 953–961.

Giasson, E., Clarke, R. T., Inda Junior, A. V., Merten, G. H., & Tornquist, C. G. (2006). Digital soil mapping using multiple logistic regression on terrain parameters in southern Brazil. *Scientia Agricola, 63*(3), 262–268.

Giraldo, M. A., Bosch, D., Madden, M., Usery, L., & Kvien, C. (2008). Landscape complexity and soil moisture variation in south Georgia, USA, for remote sensing applications. *Journal of Hydrology, 357*(3–4), 405–420.

Goel, A., & Pal, M. (2009). Application of support vector machines in scour prediction on grade-control structures. *Engineering Applications of Artificial Intelligence, 22*(2), 216–223.

Gomi, T., Sidle, R. C., Miyata, S., Kosugi, K., & Onda, Y. (2008). Dynamic runoff connectivity of overland flow on steep forested hillslopes: Scale effects and runoff transfer. *Water Resources Research, 44*(8), 1–16.

Gorthi, S. (2011). *Prediction models for estimation of moisture content*. Logan, Utah: Utah State University.

Gray, R. (2017, June 29). How can we manage the Earth's land. *BBC News*.

Haboudane, D., Miller, J. R., Pattey, E., Zarco-Tejada, P. J., & Strachan, I. B. (2004). Hyperspectral vegetation indices and novel algorithms for predicting green LAI of crop canopies: Modeling and validation in the context of precision agriculture. *Remote Sensing of Environment, 90*(3), 337–352.

Hadjimitsis, D. G., Clayton, C. R. I., & Hope, V. S. (2004). An assessment of the effectiveness of atmospheric correction algorithms through the remote sensing of some reservoirs. *International Journal of Remote Sensing, 25*(18), 3651–3674.

Hallikainen, M. T., Ulaby, F. T., Dobson, M. C., El-Rayes, M. A, & Wu, L.-K. (1985). Microwave dielectric behavior of wet soil-part I: Empirical models and experimental observations. *IEEE Transactions on Geoscience and Remote Sensing, GE-23*(1), 25–34.

Holah, N., Baghdadi, N., Zribi, M., Bruand, A., & King, C. (2005). Potential of ASAR/ENVISAT for the characterisation of soil surface parameters over bare agricultural fields. *Remote Sensing of Environment, 96*, 78–86.

Jackson, T. J., & Schmugge, T. J. (1991). Vegetation effects on the microwave emission of soils. *Remote Sensing of Environment, 36*(3), 203–212.

Jafari, A., Ayoubi, S., Khademi, H., Finke, P. A., & Toomanian, N. (2013). Selection of a taxonomic level for soil mapping using diversity and map purity indices: A case study from an Iranian arid region. *Geomorphology, 201*, 86–97.

Khanna, S., Palacios-Orueta, A., Whiting, M. L., Ustin, S. L., Riaño, D., & Litago, J. (2007). Development of angle indexes for soil moisture estimation, dry matter detection and land-cover discrimination. *Remote Sensing of Environment, 109*(2), 154–165.

Körschens, M. (2002). Importance of soil organic matter (SOM) for biomass production and environment (a review). *Archives of Agronomy and Soil Science, 48*(2), 89–94.

Kravchenko, A. N. (2008). Mapping of soil drainage classes using topographical data and soil electrical conductivity. *Handbook of Agricultural Geophysics, 1*, 255–261.

Kuusk, A. (1995). A fast, invertible canopy reflectance model. *Remote Sensing of Environment, 51*(3), 342–350.

Li, X., & Strahler, A. H. (1985). Geometric-optical modeling of a conifer forest canopy. *IEEE Transactions on Geoscience and Remote Sensing, GE-23*(5), 705–721.

Li, X., & Strahler, A. H. (1992). Geometric-optical bidirectional reflectance modeling of the discrete crown vegetation canopy: Effect of crown shape and mutual shadowing. *IEEE Transactions on Geoscience and Remote Sensing, 30*(2), 276–292.

Little, K. M., Metelerkamp, B., & Smith, C. W. (1998). A comparison of three methods of soil water content determination. *South African Journal of Plant and Soil, 15*(2), 80–89.

Long, J., Lawrence, R., Greenwood, M. C., Marshall, L., & Miller, P. R. (2013). Object-oriented crop classification using multitemporal ETM+ SLC-off imagery and random forest. *GIScience & Remote Sensing, 50*(4), 418–436.

Lu, H., Koike, T., Fujii, H., Ohta, T., & Tamagawa, K. (2009). Development of a physically-based soil moisture retrieval algorithm for spaceborne passive microwave radiometers and its application to AMSR-E. *Journal of The Remote Sensing Society of Japan, 29*(1), 253–262.

MathWorks. (2017). Introducing machine learning. In *Machine learning with MATLAB* (pp. 1–10).

McBratney, A. B., Mendonça Santos, M. L., & Minasny, B. (2003). On digital soil mapping. *Geoderma, 117*(1–2), 3–52.

Mohler, C. L., & Johnson, S. E. (2009). *Crop rotation on organic farms*. Natural Resource, Agriculture and Engineering Service (NRAES).

Oltra-Carrió, R., Baup, F., Fabre, S., Fieuzal, R., & Briottet, X. (2015). Improvement of soil moisture retrieval from hyperspectral VNIR-SWIR data using clay content information: From laboratory to field experiments. *Remote Sensing*.

Palacios-Orueta, A., Khanna, S., & Litago, J. (2005). Assessment of NDVI and NDWI spectral indices using MODIS time series analysis and development of a new spectral index based on MODIS shortwave infrared bands. In *1st International Conference of Remote Sensing and Geoinformation Processing*. Trier, Germany.

Poggio, L., Gimona, A., & Brewer, M. J. (2013). Regional scale mapping of soil properties and their uncertainty with a large number of satellite-derived covariates. *Geoderma, 209–210*, 1–14.

Prasannakumar, V., Vijith, H., Abinod, S., & Geetha, N. (2012). Estimation of soil erosion risk within a small mountainous sub-watershed in Kerala, India, using Revised Universal Soil Loss Equation (RUSLE) and geo-information technology. *Geoscience Frontiers, 3*(2), 209–215.

Schoonover, J. E., & Crim, J. F. (2015). An introduction to soil concepts and the role of soils in watershed management. *Journal of Contemporary Water Research & Education, 154*(1), 21–47.

Schrott, L., & Sass, O. (2008). Application of field geophysics in geomorphology: Advances and limitations exemplified by case studies. *Geomorphology, 93*(1–2), 55–73.

Seneviratne, S. I., Corti, T., Davin, E. L., Hirschi, M., Jaeger, E. B., Lehner, I., Orlowsky, B., & Teuling, A. J. (2010). Investigating soil moisture-climate interactions in a changing climate: A review. *Earth-Science Reviews, 99* (3–4), 125–161.

Shi, J., Wang, J., Hsu, A. Y., O'Neill, P. E., & Engman, E. T. (1997). Estimation of bare surface soil moisture and surface roughness parameter using L-band SAR image data. *IEEE Transactions on Geoscience and Remote Sensing, 35*(5), 1254–1266.

Sonobe, R., Tani, H., & Wang, X. (2017). An experimental comparison between KELM and CART for crop classification using Landsat-8 OLI data. *Geocarto International, 32*(2), 128–138.

Sutton, C. D. (2004). Classification and regression trees, bagging, and boosting. *Handbook of Statistics, 24*(4), 303–329.

Taghizadeh-Mehrjardi, R., Sarmadian, F., Minasny, B., Triantafilis, J., & Omid, M. (2014). Digital mapping of soil classes using decision tree and auxiliary data in the Ardakan region Iran. *Arid Land Research and Management, 28*(2), 147–168.

Ulaby, F., & Batlivala, P. (1976). Optimum radar parameters for mapping soil moisture. *IEEE Transactions on Geoscience Electronics, 14*(2), 81–93.

Ulaby, F. T., Razani, M., & Dobson, M. C. (1983). Effects of vegetation cover on the microwave radiometric sensitivity to soil moisture. *IEEE Transactions on Geoscience and Remote Sensing, GE-21*(1), 51–61.

Vaysse, K., & Lagacherie, P. (2015). Evaluating digital soil mapping approaches for mapping GlobalSoilMap soil properties from legacy data in Languedoc-Roussillon (France). *Geoderma Regional, 4*, 20–30.

Vishwanath, V., Sulyok, M., Weingart, G., Kluger, B., Täubel, M., Mayer, S., et al. (2011). Evaluation of settled floor dust for the presence of microbial metabolites and volatile anthropogenic chemicals in indoor environments by LC–MS/MS and GC–MS methods. *Talanta, 85*(4), 2027–2038.

Wang, A. P. (1987). Correction of atmospheric effects on remote sensing (new mathematical models and approximation solutions). *Mathematical Modelling, 9*(2), 117–124.

Wang, L., & Qu, J. J. (2009). Satellite remote sensing applications for surface soil moisture monitoring: A review. *Frontiers of Earth Science in China, 3*(2), 237–247.

Whiting, M. L., Ustin, S. L., Zarco-Tejada, P., Palacios-Orueta, A., & Vanderbilt, V. C. (2006). Hyperspectral mapping of crop and soils for precision agriculture. In *Remote Sensing and Modeling of Ecosystems for Sustainability III, Proceedings of SPIE* (Vol. 6298), 62980B-62980B–15.

Witten, I. H., & Frank, E. (2005). *Data mining: Practical machine learning tools and techniques (Google eBook)* (2nd ed.). Complementary literature None. San Francisco: Elsevier.

Yemefack, M. (2005). *Modelling and monitoring soil and land use dynamics within shifting agricultural landscape mosaic systems in southern cameroon*. The Netherlands: Utrecht University.

Zhang, Q., Sun, Z.-X., & Wang, S. (2011). Analysis of variation egularity of land-surface physical quantities over the Dingxi region of the Loess Plateau. *Chinese Journal of Geophysics, 54*(4), 436–447.

Zhang, R., Tian, J., Su, H., Sun, X., Chen, S., & Xia, J. (2008). Two improvements of an operational two-layer model for terrestrial surface heat flux retrieval. *Sensors, 8*(10), 6165–6187.

Chapter 2
Different Approaches on Digital Mapping of Soil-Landscape Parameters

In soil-landscape parameters mapping, the implementation of geomatics-GIS, GPS, remote sensing, and DEM, suggests new alternatives. Different approaches have been applied for retrieval of soil-landscape parameters. In recent years, machine learning algorithms have received increasing attention for digital mapping. Machine learning may be defined as the automated process to get the unknown relationship between input (predictor variable) and target parameters using a known example (training data), where the relationship may then be used on new data (testing data) for the purposes of prediction. The targets can be nominal or numerical. If the targets are nominal, then the problem becomes a classification one while if the target is numerical, the problem is a regression one. The learning task that involves solving such a regression problem is called supervised learning. Many predictive models also produce uncertainty maps which can focus future field activities and give the user further information about the map. Ongoing research in digital mapping has demonstrated that reasonably accurate maps can be produced using quantitative predictive models (Lagacherie et al. 1995; Burrough et al. 1997; Zhu 2000; Zhu et al. 2001; Moran and Bui 2002; McBratney et al. 2003; Scull et al. 2005; Balkovič et al. 2007). However, the quantitative modelling for prediction may expedite mapping, but complex and interconnected linkages between soil-landscape parameters under a range of soil and climatic conditions are a key open question for the scientific community. Furthermore, one of the factors that regulate the performance of machine learning algorithms is predictor variable's selection and it is a critical task.

2.1 State of the Art in Digital Soil Mapping (DSM)

Soil survey provides relevant information on the types of the soils and their distribution. Although the soil resource of an area can be assessed by carrying out the soil survey in the traditional manners. Still, it is handicapped due to the following

P. K. Garg et al., *Digital Mapping of Soil Landscape Parameters*, Studies in Big Data 72,
https://doi.org/10.1007/978-981-15-3238-2_2

facts, (i) It cannot be undertaken in the inaccessible tracts, (ii) It cannot be undertaken throughout the year, (iii) It is time-consuming. All these disadvantages of the traditional methods of soil survey can be overcomes at least to some extent by using remote sentencing technique (Reddy et al. 1993a, b; Krishan et al. 2009; Beg et al. 2011; Bhakar et al. 2012; Bansal et al. 2016; Kale et al. 2016). The application of remote sensing techniques, such as Aerial-photo-interpretation, satellite imagery interpretation, and computer analysis in the soil survey, have proved their merit over the conventional soil survey by wealth of pictorial details available and advantages arising from the three-dimensional model from stereoscopic study of the aerial photographs, contrasting surficial features from the satellite imagery and digital images (Bui and Moran 2003; Carre et al. 2007; Finke 2012; Taghizadeh-Mehrjardi et al. 2014). Other merits can be summarized in following points, (i) It is possible to systematically analyse the whole area where soil survey is to be undertaken, (ii) The fieldwork is planned in advance because the soil surveyor knows where he would take the observations and what landform he would see, (iii) The cost of survey is low, (iv) Less time is required than traditional method.

Furthermore, the spatial distribution of soil characteristics as represented by a conventional soil map does not reflect the distribution in the field because of the polygon-based mapping employed (Carre et al. 2007; Andrade et al. 2010). In polygon-based mapping, soils are assumed to be homogeneous within the polygon, and abrupt changes take place at the boundaries between polygons (McBratney et al. 2003; Taghizadeh-Mehrjardi et al. 2014). Soils usually show a diffuse spatial distribution that is hard to address in polygon-based soil maps (Triantafilis et al. 2012). Many researchers have suggested continuous raster maps as a better alternative to mapping soils and their properties. Traditional soil maps illustrate conceptual models of soil distribution on the landscape. When a soil scientist draws a line on the map he/she is predicting that certain soils are likely to be found within the delineated polygon. The parameters of this conceptual model are complicated, and are related to the individual experience or tacit knowledge of the soil scientist, and are, therefore, subjective (Zhu et al. 2001; Chen et al. 2013). Traditional soil survey methods, while thoroughly reviewed, are not assessed for accuracy.

DSM is the creation and the population of a geographically referenced soil database, generated at a given resolution by using field and laboratory observation methods coupled with environmental data through quantitative relationships. DSM has advantages over conventional soil mapping approaches as it may better capture observed spatial variability and reduce the need to aggregate soil types based on a set mapping scale (Zhu et al. 2001) (Fig. 2.1).

Hans Jenny (1941) presented theoretical framework by introducing a function to explain the current state of the soil, i.e. $S = f(\mathrm{c, o, r, p, t})$. Simply said, soil ($S$) is a function of five environmental factors: climate (c), organisms (o), relief (r), parent material (p), and time (t) (Jenny 1941). Based on Jenny's soil-forming factors (1941), McBratney et al. (2003) proposed an empirical formulation to quantitatively find correlations between spatially explicit data and the soil; Soil class function, $S = f$ (s, c, o, r, p, a, n) is referred as 'scorpan' model (McBratney et al. 2003). Two additional scorpan covariates are specifically directed towards spatial predictive

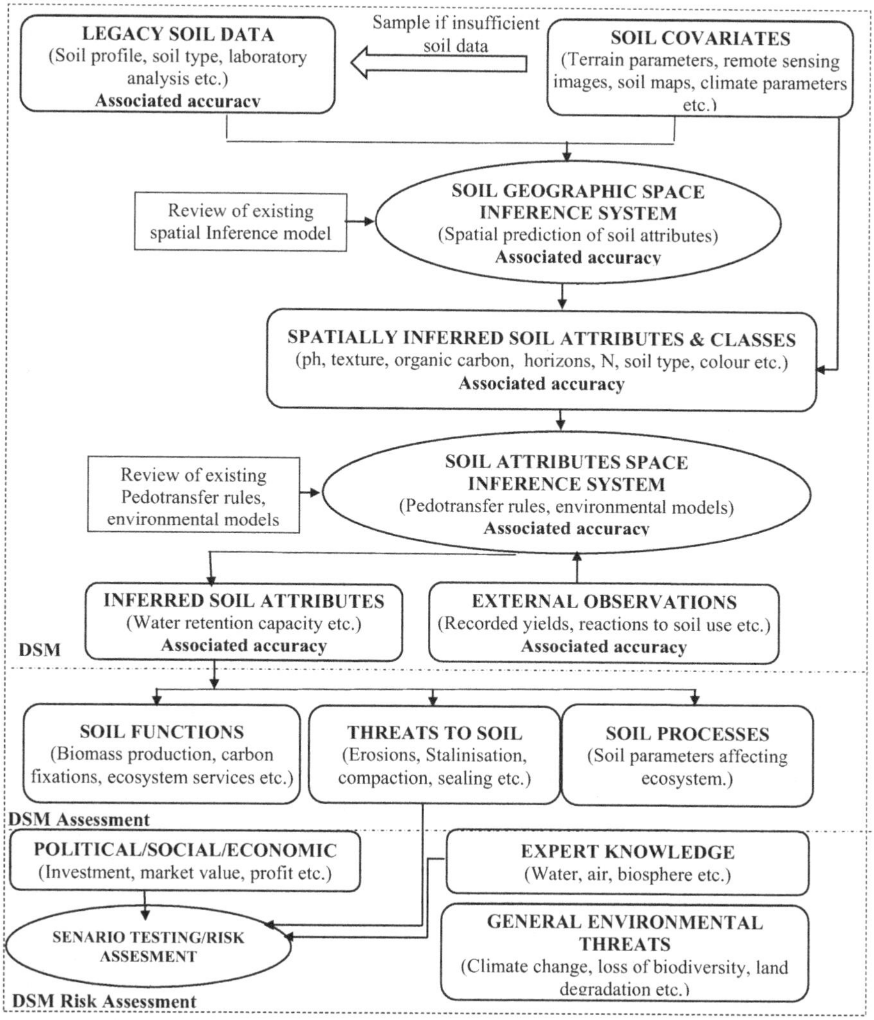

Fig. 2.1 Structure of digital soil mapping and digital soil assessment systems (Carre et al. 2007)

models: 's' soil or soil properties, and 'n' space, spatial position, or relative position. The scorpan model is a spatial framework for DSM (McBratney et al. 2003, Stum 2010). Different researchers in DSM have demonstrated that reasonably accurate soil maps can be produced (Korolyuk and Shcherbenko 2007; Connolly and Holden 2009; Panagos et al. 2011; Pásztor et al. 2012; Taghizadeh-Mehrjardi et al. 2014; Taghizadeh-Mehrjardi 2015). DSM may also expedite soil survey (Moran and Bui 2002; McBratney et al. 2003; Scull et al. 2005). Key components of DSM are the 'method' and the set of 'environmental covariates' used to predict soil classes.

A lot of research work in this field can be found using expert systems, unsupervised, and machine learning. From the last 10 years, machine learning algorithms have received increasing attention for DSM. Several machine learning algorithms such as LR, CDT, RF, ANN, and SVM techniques are investigated for DSM. With the availability of a range of satellite and attribute data scope of DSM is continuously increasing.

LR: The Hengl et al. (2007) compares semi-automated interpolation methods to produce soil class maps from profile observations and by using multiple auxiliary predictors such as terrain parameters, remote sensing indices and similar. The Soil Profile Database of Iran, consisting of 4250 profiles, was used to test different soil class interpolators. The predictors were 6 terrain parameters, 11 MODIS-EVI (Moderate Resolution Imaging Spectroradiometer-Enhanced Vegetation Index) images, and 17 physiographic regions (polygon map) of Iran. Four techniques were considered in this study, i.e. supervised classification using maximum likelihoods, MLR, regression-kriging on memberships, and classification of taxonomic distances. For both mapping of texture classes and soil types, the best prediction was achieved using regression-kriging of indicators/memberships ($\kappa = 45\%$, $\kappa = 54\%$). In all cases, kappa was smaller than 60%, which can be explained by the preferential sampling plan, the poor definition of soil classes and the high variability of soils (Hengl et al. 2007).

Marchetti et al. (2011) used MLR to produce predicted soil maps. Relationships between soil type distribution and terrain attributes—slope, incoming solar radiation, NDVI, TWI, etc.—were established, and the most significant were used in an MLR to generate simulated maps. These maps, derived from a set of measured point data, auxiliary information from a Digital Terrain Model and Landsat images, were compared with the soil subsystem map 1:250,000, realized by ARSSA (the Regional Agency for Agricultural Extension Services of Abruzzo Region). Different scenarios were built, and the best was compared with a traditional soil map. Agreement among maps, but simulations better represent the local morphology. MLR produced explicit and objective maps with a measurable level of accuracy. MLR allows to decrease costs and subjectivity of soil maps (Marchetti et al. 2011). MLR has been used successfully to estimate specific soil characteristics rather than soil taxonomic classes (Campling et al. 2002; Ohlmacher and Davis 2003; Kravchenko 2008) but since soil map units are categorical variables, the MLR may be suitable for predicting the occurrence of soil classes, providing the degree of probability (Giasson et al. 2006).

Jafari et al. (2012) compared binary logistic regression as an indirect approach and MLR as a direct approach to produce soil class maps in the Zarand region of southeast Iran. Among the predictors, the geomorphology map was identified as an important tool for DSM approaches as it helped to increase the accuracy. The results of prediction showed larger mean probability values for each great soil group in the areas actually covered by the great soil groups compared with other areas, indicating the reliability of the prediction (Jafari et al. 2012, 2013). The generalized linear geostatistical model was presented by Kempen et al. (2012), for soil type mapping. The model is applied to map the soil in a peatland area using real-world data. Predictive soil map was validated with independent probability sample data. The map was not more accurate than soil maps obtained by non-spatial models (Kempen et al. 2012).

CDT: The CDT technique has been used to create a map of soil class for an expanded area from a small reference area (Lagacherie et al. 1995; Cialella et al. 1997; McKenzie and Ryan 1999), and to improve an existing soil map (Moran and Bui 2002). Cialella et al. (1997) used DTA to predict soil drainage class from remotely sensed and digital elevation data, and McKensie and Ryan (1999) used regression trees to model individual soil properties.

Bui et al. (1999) established that soil mapping rules could be derived from an existing 1:100 000 surveys of soil associations in the Toowoomba area using a combination of a 9-s DEM, derived terrain attributes, and geology. The assumptions underlying rule development were (Bui et al. 1999): (1) that soil distribution reflects the long-term interactions between terrain variables, geology, and vegetation in landscapes, (2) that a DEM and derived terrain attributes can represent factors of soil formation digitally, and (3) that the existing soil maps have captured soil–landscape interactions in the area mapped.

Moran and Bui (2002) improved the predictive capability of the rules in the Toowoomba area by adding as predictor variables remotely sensed data (Landsat Multi Spectral Scanner: Landsat MSS), different terrain attributes (uphill distance from nearest river and downhill distance from nearest hilltop), and adaptively filtered data to retain local spatial information (Moran and Bui 2002).

Bui and Moran (2003) collated existing digital soil maps and used inductive spatial modelling to predict soil types from those maps combined with environmental predictor variables. Lithology, Landsat MSS, the DEM of Australia, and derived terrain attributes, all gridded to 250-m pixels, were the predictor variables. Rules for the spatial distribution of soils were derived from classification tree models.

In Scull et al. (2005) study, a GIS was used to manage the digital spatial data characterizing the environmental variables that are related to, or surrogates for, the soil-forming processes. This study utilized four sources of data as input to the GIS: remote sensing, terrain analysis, pre-existing environmental maps (lithology, climate, a generalized statewide soil database, and vegetation), and soil survey data. Decision tree modelling approach was developed using soil data as the dependent variable and environmental data from the GIS as independent variables. The study suggested that soil taxonomic class can be predicted with reasonable accuracy from environmental variables.

Kim et al. (2012) also used CDT technique to predict soil series using spectral data and ancillary environmental data sets. This study compared the effects of spatial resolution (10, 30, and 250 m) by using three satellite images acquired (i) satellite pour l'observation de la terre, i.e. Satellite for observation of Earth (SPOT, 10 m), (ii) Landsat Enhanced Thematic Mapper Plus (ETM+, 30 m), and (iii) MODIS, 250 m. The soil series prediction model derived using SPOT spectral data 'without bedrock depth' and the model derived using MODIS spectral data 'with bedrock depth' showed the best results based on accuracy measures and Kappa coefficient.

Results of this study suggested that the variability of soil series can be explained by bedrock/parent material > topographic variables > vegetation properties derived from remote sensing (Kim et al. 2012).

RF: Developed by Breiman (2001), RF is an ensemble of the CART. RF is said to be as accurate as or better than adaptive boosting, yet computationally faster (Breiman et al. 1984; Breiman 2001; Sutton 2004). Grimm et al. (2008) proposed RF-based digital Soil Organic Carbon (SOC) mapping approach, the study provided SOC estimates of high spatial resolution and estimates of error and predictor importance. This relies on a soil inference model based on spatially referenced environmental layers of topographic attributes, soil units, parent material, and forest history (Grimm et al. 2008). The environmental variables that explained most of the variation in the topsoil (0–10 cm) were topographic attributes. In the subsoil (10–50 cm), SOC distribution was best explained by soil texture classes as derived from soil mapping units in Grimm et al. (2008) study.

The Poggio et al. (2013) aimed to test the potential of time series of MODIS vegetation and drought indices to provide relevant information to model topsoil properties in a Boreal-Atlantic region (Scotland) focussing on the differentiation between soils with high and soils with low organic matter contents. Terrain parameters derived from Shuttle Radar Topography Mission (SRTM) were also included. Two approaches were used in the modelling to deal with both continuous and categorical predictands: General Additive Models (GAMs) and RF. RF was used to predict organic matter classes and the presence/absence of organic soils. In order to integrate the information from the GAM in the RF, the probability predictions resulting from the GAM, for the considered soil property, were included in the pool of covariates used for the RF (Poggio et al. 2013). The use of MODIS data as covariates increased the predictive capabilities of GAM using only terrain parameters. The misclassification error for organic matter classes was between 25 and 35%.

Pahlavan Rad et al. (2014) investigated the use of conditioned Latin hypercube sampling and RF for the mapping Soil Taxonomy great group, subgroup, and series levels in Golestan Province, Iran. The overall error was 48.5, 51.5, and 56.6% for a great group, subgroup, and series levels, respectively. Estimated individual soil type error was between 8 and 100%. Soil types with larger sample sizes were predicted over a greater area at each taxonomic level. The soil adjusted vegetation index, the conventional soil series map, and geomorphology were the most important covariates for each taxonomic level. Taxonomic classes with important covariates had low OOB error. The updated soil series map was 13.4% more accurate than the existing conventional soil series map (Pahlavan Rad et al. 2014).

Vaysse and Lagacherie (2015) study used RF and RF plus kriging to provided estimates of 29 soil properties over a 100 m by 100 m grid, i.e. clay, sand, silt, coarse fragment, organic carbon contents, pH and CEC, at the four upper GSM (Global Soil Map)-specified depth intervals, plus soil depth. Four DSM models were tested to map 29 GSM-specified soil properties in Southern France. Mapping performances were highly variable (from $R^2 = 0$ to $R^2 = 0.79$) (Vaysse and Lagacherie 2015). Soil profile-based DSM models generally outperformed soil map-based DSM models.

The soil database used as an input of DSM models was the most limiting factor. Bagging trees and RF approaches have the best performances among CART methods (Prasad et al. 2006). While RFs have often been used in ecological modelling, few examples exist in DSM (Grimm et al. 2008; Vaysse and Lagacherie 2015).

ANN: Moonjun et al. (2010) employed ANN and DT within the GIS environment to comply with the complexity of the soil-forming factors governing the soil formation. In this study geopedologic approach to soil survey, which was based on an understanding of the landscape-soil relationship, was helpful to obtain spatial soil information in inaccessible areas. The study further suggested that ANN and DT are useful techniques in modelling the complex interactions among the soil-forming factors. The difference, however, is that ANN, once it is well learned, is faster, thus more recommendable in terms of time and cost-saving (Moonjun et al. 2010).

Jafari et al. (2013) discussed the selection of the taxonomic level for soil mapping in an arid region in southeast Iran on the basis of (1) the purity of a digital soil class map derived from an Artificial Neural Network (ANN) prediction method using environmental covariates and (2) pedodiversity indices of these soil maps. In most predictions, the combined use of terrain attributes and geomorphic surfaces provided the best results (Jafari et al. 2013). This study concluded that when the taxonomic level changed from order to subgroup, the purity decreased, whereas the values of the diversity indices increased. The highest purity and lowest diversity were observed at the order level, indicating a good quality map in terms of its purity, but reflecting only a little soil diversity, thus with a low usage potential.

In a recent study, Bagheri Bodaghabadi et al. (2015), used multilayer perceptron Artificial Neural Networks (ANN) to map soil units using DEM attributes. Several optimal ANN were produced based on a number of input data and hidden units. Training errors were below 11% for all the ANN models applied, while the test error (interpolation error) and validation error (extrapolation error) were as high as 50% and 70%, respectively. As expected, soil prediction using a higher level of classification presented a better overall level of accuracy. To obtain better predictions, in addition to DEM attributes, data related to landforms and/or lithology as soil-forming factors, should be used as ANN input data (Bagheri Bodaghabadi et al. 2015).

SVM: The study Kovačević et al. (2010) introduced the application of SVM in the estimate of values of soil properties and soil type classification based on known values of particular chemical and physical properties in sampled profiles. Comparison of the proposed approach with other LR models showed that SVM is the model of choice for estimation of values of physical properties and pH value when using only chemical data inputs. They are also the model of choice in the cases where chemical data inputs are not strongly correlated to the estimated property. However, in classification task, their performance was similar to that of the other compared methods, with an increasing advantage when a data set consists of a small number of training samples per each soil type (Table 2.1).

DSM approach is still unexplored in the Indian context. However, the Indian government is eager to create digital maps of soil type and moved one step forward

Table 2.1 Summary of few research works on DSM

Study	Model	Predictors					Predicted factor
		s	c	o	r	p	
Bell et al. (2000)	Linea Regression (LR)	Soil map		Aerial photograph	DEM		Soil organic corban content
Bourennane et al. (2000)	Kriging with external drift				DEM		Soil horizon thickness
Chaplot et al. (2000)	Kriging				Elevation above stream bank slope, specific catchment area, CTI		Hydromorphic Index
Dobos et al. (2000)	GLM				Elevation, slope, PDD		Soil classes
Odeh and McBratney (2000)	GLM, GAM, RT, ANN, Kriging, co-kriging, Regression-kriging	ECa		Crop yield	DEM		Clay content, CEC
Zhu (2000) Bishop and McBratney (2001)	GAM, RT, Kriging, CDT	Soil ECa, aerial photograph, Landsat TM		Crop yield	DEM		CEC
Bui and Moran (2001)	CDT	Soil map, Landsat TM			DEM	Lithology map	Soil classes
Carré and Girard (2002)	ANN	Soil map		SPOT	DEM	Geological map	Soil classes
McBratney et al. (2003)	ANN				DEM	Stratigraphic geology map	Soil classes
Sommer et al. (2003)	Rule based	EM induction survey		Airborne multispectral scanner	DEM, curvature and upslope catchment area		Soil classes
Scull et al. (2005)	CDT	Generalized state wide soil database	Climate maps	Land use map	DEM	Lithology map	Soil taxonomic class

(continued)

Table 2.1 (continued)

Study	Model	Predictors					Predicted factor
		s	c	o	r	p	
Hengl et al. (2007)	Supervised classification using maximum likelihoods, MLR		Climate data		DEM		Soil classes
Grimm et al. (2008)	RF	Soil map		Aerial photographs	DEM	Geological map	Digital Soil Organic Carbon (SOC)
Kovačević et al. (2010)	SVM, LR						Soil properties, Soil classes
Moonjun et al. (2010)	ANN and DT		15 years of climate data	Forest reports	Aerial photographs, Ortho-photo	Lithology map	Soil classes
Marchetti et al. (2011)	MLR	Landsat 7 TM, Grain Size Index (GSI)		NDVI, TWI	DEM, curvature profile and plan, incoming solar radiation, TWI		Soil class
Kim et al. (2012)	CDT	Field observations		EVI, Moisture Stress Index (MSI), NDVI, Simple Ratio (SR), Transformed Vegetation Index (TVI), Normalized Difference Water Index (NDWI)	USGS, High accuracy elevation data set	Soils Map	Soil series map

(continued)

Table 2.1 (continued)

Study	Model	Predictors					Predicted factor
		s	c	o	r	p	
Jafari et al. (2013)	ANN			NDVI, Ratio Vegetation Index (RVI), Perpendicular Vegetation Index (PVI), Clay Index (CI)	ASTER-GDEM	Geomorphological map, Geological map	Soil classes
Pahlavan Rad et al. (2014)	RF			Soil Adjusted Vegetation Index (SAVI)	Aster-GDEM derived aspect, maximum downward slope, downslope flow-path length, mean upslope slope, curvature, sediment transport capacity index, slope, TWI	Soil series map, Geomorphology	Soil series map
Bagheri Bodaghabadi et al. (2015)	Multilayer perceptron ANN				DEM, slope, aspect, curvature, maximum and minimum curvature, curvature		Soil classes

with The Bill & Melinda Gates Foundation (BMGF) to generate digital maps of soil type and properties of one of its states, Andhra Pradesh (The Hindu, November 16, 2017). It is a pioneering step in soil mapping in India. DSM and space techniques can accelerate the provision of fine-scale information on soil fertility and conditions in India. It can also be used in monitoring and understanding soil nutrient change over time.

2.2 Machine Learning in Crop Cover Mapping

Though several studies reported the use of different image processing techniques for land use/cover classification, there are only a few studies utilizing ANN for agricultural crops inventory. ANN has shown promise in the areas of pattern recognition and classification (Murthy et al. 2003; Paloscia et al. 2008; Rodriguez-Galiano and Chica-Rivas 2014; Xin et al. 2015). Murthy et al. (2003) evaluate the performance of Maximum Likelihood Classifier (MLC) and ANN with a backpropagation method for a wheat crop with multi-temporal images. In this study, the performance of ANN was proved to be better, in both the situations of dominant wheat and less dominant wheat cultivation (Murthy et al. 2003). Consequently, the application of multidate or multi-temporal image data has been recommended to improve the discrimination ability (Ravan et al. 1995, 2004; Roy and Ravan 1996; Ravan and Roy 1997; Boloorani et al. 2008; Fang et al. 2013; Chaudhary et al. 2016a, b). The multi-temporal classifications provide additional information on the phenology of vegetation to the classification process and, as a result, the classification possibilities would be notably enhanced (Young 2003; Murthy et al. 2003; Young and Harris 2005; Jeyaseelan et al. 2007; Krishan et al. 2009; Seneviratne et al. 2010; Bhakar et al. 2012; Blahwar et al. 2012; Kappas 2013; Pandey et al. 2015; Ranjan et al. 2015; Wang et al. 2016; Mondal et al. 2016; Reddy et al. 2017; Watham et al. 2017; Kothapalli Venkata et al. 2017).

ANN: One of the problems of agriculture in India is that in many zones of the country crops are grown under a wide range of soil and climatic conditions that are not necessarily ideal for a particular crop (Patel et al. 2006; Srivastava et al. 2008, 2013). The neuro-fuzzy technique can also be used to suggest alternative crops by integration of multisource data (Sarkar et al. 2008). Sarkar et al. (2008) used elevation, rainfall, and goodness values of different existing crops as input. The elevation and rainfall were determined from contour information of land-survey maps. The other components, such as goodness values of different existing crops, were based on the productivity estimates of soil determined by fuzzyfication and expert opinion (on soil) along with production quality by the NDVI obtained from satellite imagery (Sarkar et al. 2008). Finally, for each pixel an alternative crop suggestion has been obtained with the ANN method. Since the 1980s, the application of ANN in crop identification using remote sensing images has achieved great progress. However, because of the inherent disadvantage of neural network theory, there are problems such as overfitting, the difficulty of mechanism analysis, and structural instability.

To improve the accuracy, timeliness, and effectiveness of crop identification SVM was also used by many researchers (An and Yang 2007; Mathur and Foody 2008; Tan et al. 2011; ChenChi et al. 2012; Sonobe et al. 2014a).

SVM: An and Yang (2007) developed multi-crop identification model, based on Stepwise Removal learning-Support Vector machine (SR-SVM) using remote sensing images. This classifier was constructed using a Binary Tree (BT) structure. In each layer, it applies the SR learning algorithm with SVM to a classification which reduces the calculation complexity and sample training time. In this study, two vegetation indices including the NDVI and the Ratio Vegetation Index (RVI) were used. Integrating the idea of SR, an optimal solution can be found so that the iterative error in the hierarchical method can be eliminated (An and Yang 2007).

The accuracy of supervised classification, like SVM, is dependent to a large extent on the training data used (Pal and Mather 2003a, b, 2005; Pal 2008). An approach to identify informative training samples was demonstrated by Mathur and Foody (2008) for the classification of agricultural classes in southwestern part of Punjab state, India. With the intelligent training scheme, training data were, therefore, acquired from sites with relatively extreme spectral responses (potential border training samples) that should provide appropriate support vector. Although the intelligent training approach requires some knowledge-based input, this study was based on the impact of well-known variables such as water content, growth stage, and soil background on the spectral response of vegetation. Relative to the use of a conventional scheme, the accuracy of the resulting classification was 1.34% less accurate at 90.66% but involved substantially less effort and could be acquired at less expense (Mathur and Foody 2008).

Tan et al. (2011) present the application of a hybrid classification technique of Entropy Decomposition and Support Vector Machine (EDSVM) for crop-type categorization. It takes the advantage of the desired parameters from the Entropy Decomposition (ED) method and the SVM method that determines the optimal separation between classes in a higher dimensional feature space to improve on the existing classification results. In this study, EDSVM classifier was applied on a multi-crop field Airborne Synthetic Aperture Radar (AIRSAR) image of Flevoland in the Netherlands and the robustness of the classifier was evaluated. At the same time, a hybrid Entropy Decomposition and Neural Network (EDNN) classifier method was developed to validate the effectiveness of the EDSVM classifier. As a result, EDSVM was proved to be robust and to yield a superior result compared with ANN, SVM, and EDNN classifiers in Tan et al. (2011) study.

ChenChi et al. (2012) investigated rice cropping practices and rice-growing areas in the Vietnamese Mekong Delta using MODIS 250 × 250 m NDVI data acquired during the 2002 and 2007 rice cropping seasons. The methodology involved extraction and evaluation of phenological rice training pattern from the smooth time profiles of NDVI and classification of rice cropping using SVM. At the district level, there was good agreement between the MODIS-based estimated areas and government rice statistics for 2002 and 2007 ($R^2 \geq 0.85$) (ChenChi et al. 2012). Many traditional machine learning algorithms generate a single model (e.g. a decision tree or neural network). Ensemble methods have been called the most influential development in data mining and machine learning in the past decade (Weigend et al. 1995; Tumer and Oza 2003). Better predictive performance could be obtained from ensembles than any of the constituent learning algorithms (Dietterich 2000; Rätsch et al. 2001; Punera and Ghosh 2008).

RF: Deschamps et al. (2012) compared DT with RF for crop-type classification using radar data. Overall classification accuracies were improved by approximately 4–5% over traditional boosted DT with gains of up to 7% in the accuracies of specific classes. Accuracies above 85% were obtained for key crops including canola, soybeans, corn, and wheat. The HV and VV polarizations had the most significant contributions, while the HH polarization contributed little throughout the season, except in late September when the HH response was largely driven by soil conditions. The RF classifier provided large performance gains in terms of processing time relative to the decision tree classifier (Deschamps et al. 2012).

Sonobe et al. (2014a, b) used the RF classifier and compared the results with Classification And Regression Tree (CART), to evaluate the potential of multi-temporal Terra SAR-X dual-polarimetric data, on the Strip Map mode, for the classification of crop type. Furthermore, comparisons of the two algorithms and polarizations were carried out. In the study area, beans, beet, grasslands, maize, potato, and winter wheat were cultivated, and these crop types were classified using the data set acquired in 2009. The classification results of RF were superior to those of CART, and the overall accuracies were 0.91–0.93. In another study (Sonobe et al. 2014b) described the comparison of three different classification algorithms, i.e. CART, SVM, and RF for mapping crops in Hokkaido, Japan, using Terra SAR-X data. In the study area, beans, beets, grasslands, maize, potatoes, and winter wheat were cultivated. However, in this study SVM was the optimum algorithm with overall accuracy of 89.1% for the same year classification, which is the classification using the training data in 2009 to classify the test data in 2009, and 78.0% for the cross-year classification, which is the classification using the training data in 2009 to classify the data in 2012. In his recent study, an experimental comparison between KELM and CART for crop classification using Landsat-8 OLI data was also presented (Sonobe et al. 2017). The OLI data acquired on 8 July 2013 was used for crop classification of beans, beets, grassland, maize, potatoes, and winter wheat. The KELM algorithm performed better in this study and achieved overall accuracies of 90.1%. According to the Jeffries–Matusita (J–M) distances, the short wavelength infrared band provides a greater contribution (the highest value was observed for band 6 in OLI data).

Kumar et al. (2015a, b) presented a comparison of SVM, ANN, and Spectral Angle Mapper (SAM) algorithms for crop classification using Linear Imaging and Self-Scanning Sensor (LISS IV) data. The outcome of this study indicated better performance of SVM and ANN algorithms in comparison to SAM for the classification using LISS IV sensor data. The overall accuracies obtained by SVM and ANN were 93.45% and 92.32%, respectively, whereas the lower accuracy of 74.99% was achieved using the SAM algorithm through error matrix analysis. The author also evaluated SVM in comparison to maximum likelihood and normalized difference vegetation index classification schemes for crop type such as barley, wheat, lentil, mustard, pigeon pea, linseed, corn, pea, sugarcane, and other crops and non-crop such as water, sand, built up, fallow land, sparse vegetation, and dense vegetation (Kumar et al. 2016). The results were statistically analysed and compared using Z-test and χ2-test. Statistical analysis showed that the accuracy results using SVM with polynomial of degrees 5 and 6 were not significantly different and found better than the other classification algorithms (Kumar et al. 2015b).

Shukla et al. (2018) also implemented RF algorithm for crop mapping across an aridic to ustic area of Indian states. To incorporate dependency of crops in different variables, namely, texture, phenological, parent material and soil, soil moisture, topographic, vegetation, and climate, 35 digital layers are prepared using different satellite data (ALOS DEM, Landsat-8, MODIS NDVI, RISAT-1, and Sentinel-1A) and climatic data (precipitation and temperature). This study indicated the importance of spectral responses as well as of topography in model development. Using a multitude of variables from numerous data sources as input to an RF classifier, good results have been achieved. This study has shown the good capability of RF model as a crop class predictor for large areas. Furthermore, most popular classical ensemble models—bagging/ARCing, gradient boosting, Importance Sampled Learning Ensemble (ISLE), DT have been selected for comparative analysis (Shukla et al. 2018). The performance was evaluated on the basis of marginal rates, F-measure and Jaccard's coefficient of community, Classification Success Index and Agreement Coefficients. The score is calculated to rank the algorithm. This study also indicated RF as the best performer followed by gradient boosting for crop classification. Other ensemble methods ARCing, bagging, and ISLE were in decreasing order of performance. Traditional non-ensemble method decision tree scored higher than ISLE (Shukla et al. 2018). However, different data sources have been implemented and tested for modelling in different studies, nowadays availability of LIDAR data open scope for further study to derive terrain attributes (Dashora et al. 2006, 2013; Lohani et al. 2006; Lohani and Singh 2008; Ghosh and Lohani 2013).

2.3 State of the Art in Soil Moisture Estimation

Traditional approach: Tradition approaches are basically ground-based techniques. In these techniques of soil moisture estimation, the instrument is in direct contact with soil particles. Such techniques provide more precise data. The gravimetric method involves oven drying a soil sample of known volume at 105 °C for 24 h. The water content is calculated by subtracting the oven-dry weight from the initial field soil weight (Lunt et al. 2005). It seems that this temperature range has been based on water boiling temperature and does not consider the soil's physical and chemical characteristics. Time Domain Reflectometry (TDR) and Frequency Domain Reflectometry (FDR) are based on the interaction between electromagnetic wave and soil moisture. In TDR, the travel time T and the length of the probe L, which has been travelled along twice, the propagation velocity $V = 2L/T$ is calculated (Seneviratne et al. 2010). The velocity of the pulse changes with changes in soil moisture content due to the relatively large dielectric value of water (Muñoz-Carpena 2015) is the basic principle behind soil moisture measurement using TDR. The limitation of this method is the minimum system rise time. The total rise time consists of the combined rise time of the driving pulse and that of the oscilloscope which monitors the reflections. FDR is similar to TDR, but FDR provides an estimate of the soil moisture content on the basis of a variation in the frequency of a signal due to the dielectric properties of the soil (Trenberth et al. 2009). Sensitivity to air gaps, soil salinity, temperature, bulk density, and clay content limit the use of this method (Erlingsson et al. 2009).

Methods using radioactive sources for pulse generation are also used for soil moisture estimation, but due to environment aspect, these methods have limited applications (Neutron Probe Technology, Gamma Ray Attenuation). In Neutron Probe Technology, the neutrons are slowed down by collisions with the nuclei of hydrogen atoms present in the molecules of water in the soil. Since water contains two atoms of hydrogen per molecule, this therefore, gives a measure of soil moisture. The output from the neutron probe can be directly related to the soil moisture content (Chanasyk and Naeth 1996). The Gamma Ray Attenuation is also a radioactive method capable of determining the moisture content in the upper soil layers (up to 1–2 cm). The gamma transmission gives a measure of the change in wet density, which can give a direct estimate of the soil moisture content (Seneviratne et al. 2010). Another instrument, tensiometer measures the capillary tension. But tensiometers have been found to be unsuitable for measurements in dry soil. The high maintenance requirements limit the use of this method in research (Dukes et al. 2010).

The Ground Penetrating Radar (GPR) measurements are based on the transmission and reflection of an electromagnetic wave in the soil. This method bridges the gap between point measurements and remote sensing (Dobriyal et al. 2012). Steep and rocky slopes limit its use due to the large sizes of the antennas. The use of GPR in forests is difficult because trees behave as reflectors generating erroneous data (Schrott and Sass 2008). Comparison of traditional ground-based techniques has been presented in Table 2.2, on three basic scales cost, accuracy, and response time. Measured parameters limitations and drawbacks of these methods have been also listed.

Remote sensing in soil moisture estimation: As mentioned above that conventionally, in situ soil moisture measurements can be done in either direct or indirect method, as described in the above section. These methods are cost-prohibitive, time- and resource-consuming. Therefore, research oriented towards physical and empirical models. Physical models involve complicated scattering phenomena from probabilistic models of soil, vegetation, and terrain whereas empirical models capture relationships among measured variables to estimate geophysical characteristics (Ahmad et al. 2010).

With advancement in remote sensing techniques, a lot of research has been concentrated in this area based on the facts that reflected electromagnetic spectrum in the visible and infrared range is highly correlated to soil properties like soil colour, texture, surface roughness, silt, phosphorus and organic matter content, soil pH,

Table 2.2 Comparison of traditional ground-based techniques

Methods	Technique	Scales			Measured parameters	Suitability		
		Cost	Accuracy	Response time		Limitations	Drawbacks	References
Gravimetric	Oven drying and weighing	Economical	High	24 h	Mass water content or	Rocky soils	Labour and time intensive, destructive	Trenberth et al. (2009)
Neutron Probe	Neutrons hydrogen collision	Expensive	High	1–2 min	Volumetric soil moisture content	Soil specific calibration	Radioactive source hazardous to health and environment	Dobriyal et al. (2012)
TDR	Change in EM wave velocity with dielectric constant	Economical	High	Instant	Volumetric soil moisture content	High saline soils	Probe length influence accuracy	Richardson et al. (1992)
FDR	Change in EM wave frequency with dielectric constant	Expensive	Low ± 0.01	Instant	Volumetric soil moisture content	Soil specific calibration, soil salinity clay content	Sensitive to air gap, Temperature	Erlingsson et al. (2009)
Tensiometers	Suction force	Economical	High	Instant	Soil water potential	Dry soil	High maintenance	Dukes et al. (2010)
GPR	Change in reflected signature	Expensive	High	Instant	Volumetric soil moisture content	Forest area, steep and rocky slope	Soil dependency	Schrott and Sass (2008)
Gamma ray attenuation	Gamma ray absorption relation with wet density	Expensive	High	Instant	Volumetric soil moisture content	Highly stratified soil	Difficulty of use	Pires et al. (2005)

available nitrogen, vegetation composition, as well as soil moisture (Gessler et al. 1995; Iverson et al. 1997; Morris and Boerner 1998; Hutchinson et al. 1999; Boerner et al. 2000; Hulley et al. 2010; Pandey et al. 2013, 2016; Kumar 2015a, b; Kumar et al. 2015a; Chaudhary et al. 2016a, b). Table 2.3 is presenting the advantages and limitations of different remote sensing data for soil moisture study. Remote sensing measurements of the soil record the amount of radiation in a given wavelength reflected off of or emitted from the surface to the sensor. The basic difference among different approaches is the wavelength bands of electromagnetic spectrum used, the source of electromagnetic energy, response measured by the sensor and model to relate signal response to soil moisture content (Gens 2000, 2003). For example, surface temperature property is used in a thermal infrared band of electromagnetic spectrum to retrieve surface soil moisture. Besides those common to all optical techniques, such as shallow soil penetration, cloud contamination, rain effect, the utility of TM band is limited by meteorological condition and vegetation. They are often empirical and depend on local meteorological conditions, such as wind speed, air temperature, and humidity (Erasmi et al. 2004; Mallick et al. 2009; Seneviratne et al. 2010; Bauböck et al. 2014; Noi et al. 2016).

The two angle indexes, Shortwave Angle Slope Index (SASI) and Angle at NIR (ANIR) are presented by Khanna et al. (2007). This study was based on the previous research, carried out by Palacios-Orueta et al. (2005) to develop Shortwave Angle Normalized Index (SANI) with a combination of NIR, SWIR1, and SWIR2 MODIS bands. The angle index, ANIR, was a combination of reflectance values in the Red, NIR, and SWIR1 bands. Soil moisture samples regressed on soil moisture content had an average R^2 of 0.95. Thus the strong correlations of SASI and slope across all sample sites indicate that the index could be effectively applied at a regional scale, although it may not be accurate at a global scale (Palacios-Orueta et al. 2005; Khanna et al. 2007).

Giraldo et al. (2008) have examined the effect of different LULC on the temporal and spatial variation of soil moisture. In highly diverse landscapes, the uncertainty of the soil moisture estimate from remote sensors is likely to increase by using sensors at coarse spatial resolution. This analysis indicates that regardless of the spatial proximity between two homogeneous fields, LULC is a factor affecting the soil moisture behaviour. The analysis also showed high homogeneity in the soil moisture behaviour within them (Giraldo et al. 2008).

Trapezoid T_S–F_V (surface temperature–vegetation fraction cover) space (Zhang et al. 2008; Sun et al. 2012): Trapezoid T_S–F_V (surface temperature–vegetation fraction cover) space was introduced by Zhang et al. (2008) in their model, considering the dry edge and wet edge form two physical boundary of the 2D T_S–F_V space (Sun et al. 2012). Figure 2.2, shows a conceptual diagram of Trapezoid T_S–F_V (surface temperature–vegetation fraction cover) space.

An Advanced Temperature Vegetation Dryness Index (ATVDI) is used to monitor soil moisture status by Sun et al. (2012).

Table 2.3 Summary presenting the advantages and limitation of remote sensing data for near-surface soil moisture estimation

Spectrum domain	Spectrum range	Properties observed	Suitability	Advantages	Limitations
Optical	0.4–0.9 μm	Soil reflection	Vegetated area	• Fine spatial resolution • Broad coverage	• Limited surface penetration • Cloud contamination • Smoke, fog, rain effect • Weather dependency
Thermal	3–5 μm and 8–14 μm	Radiant temperature	Bare land	• Fine spatial resolution • Broad coverage	• Limited surface penetration • Cloud contamination • Rain effect • Perturbed by meteorological conditions and vegetation
Microwave					
(a) Passive	0.15–30 cm	• Brightness temperature • Dielectric constant	Both	• Weather-free operation • Moderate surface Penetration	• Coarse resolution • Sensitive to roughness and moisture in vegetation
(b) Active	1.0–100 cm	• Backscattering coefficient • Polarization • Dielectric constant	Both	• Weather-free operation • Moderate to high surface penetration (1 cm to 1 km) • High spatial resolution	• Varying resolution from near to far range • Sensitive to roughness and moisture in vegetation • Speckles

Fig. 2.2 T_S–F_V space

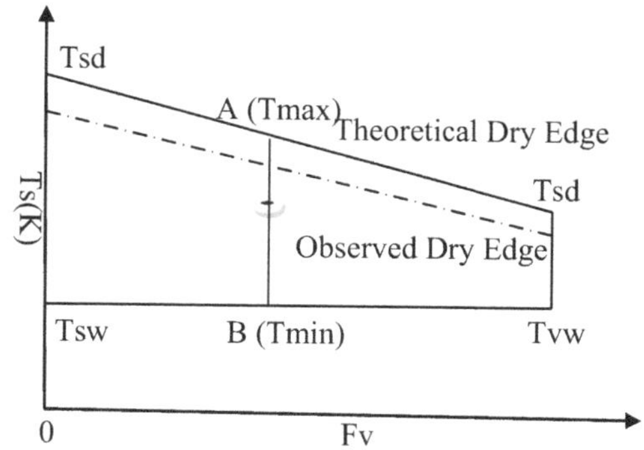

Assuming that the vegetation index is linearly related to vegetation fractional cover (Fv), Zhang et al. (2008) proposed a method to determine the theoretical dry edge and wet edge. It is found that the observed dry line, which is directly determined by the scatter plots from the remote sensing data, is usually lower than the theoretical edge (Stisen et al. 2008).

$$\text{TVDI} = 1 - \text{SWI} = \frac{T_S - T_{min}}{T_{max} - T_{min}} \tag{2.1}$$

The TVDI calculated from the theoretical dry edge is called the Advanced Temperature Vegetation Dryness Index (ATVDI) and that calculated from the observed dry edge is called the Simple Temperature Vegetation Dryness index (STVDI).

Mallick et al. (2009) demonstrated a technique to estimate volumetric surface (0.05 m) soil moisture content Ø_v using LST-NDVI 2D space in cropped soils over selected agricultural regions of India. Figure 2.3 shows a conceptual diagram of LST-NDVI triangle for determining the soil wetness index.

Triangular scatter space from Land Surface Temperature (LST) and normalized difference vegetation index was utilized to obtain a Soil Wetness Index (SWI). SWI on a given day and time (t), is defined as

$$\text{SWI(t)} = \frac{T_{S_{max(i)}} - T_{S_{(i)}}}{T_{S_{max(i)}} - T_{S_{min}}} \tag{2.2}$$

where $T_{S_{(i)}}$ is the land surface temperature of the i-th pixel. $T_{S_{min}}$ is the minimum LST in the triangle that defines the wet edge and $T_{S_{max(i)}}$ is the maximum LST corresponding to i-th NDVI along the dry edge. By knowing the upper and lower limits of volumetric soil moisture at the surface, the soil wetness index on the given day or time can be converted to an absolute estimate of volumetric surface soil moisture using the following relationship (Wagner et al. 1999; Mallick et al. 2009).

$$\text{Ø}_V = \text{Ø}_{min} + \text{SWI}(\text{Ø}_{max} - \text{Ø}_{min}) \tag{2.3}$$

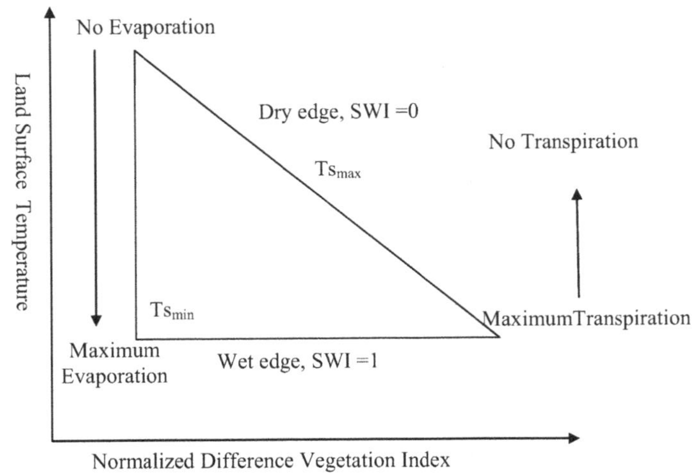

Fig. 2.3 LST-NDVI triangle space

where $Ø_{min}$ is volumetric soil moisture content at permanent wilting point and $Ø_{max}$ is arithmetic mean of volumetric soil moisture content at saturation and field capacity. Validation of field scale $Ø_v$ estimates from ASTER produces an RMSE of 0.039. This study showed that the validity of surface moisture estimates largely depends on dynamic ranges of LST and NDVI, which often may not be sufficient as a result of restricted sampling window size due to low swath in case of finer resolution sensors. The LST–NDVI triangle method showed less error for the intermediate NDVI range 0.35–0.65. This revealed a fairly good correlation (0.75) and low RMSD of 0.027 for fractional vegetation cover (Fv) below 0.5.

Transfer coefficient (ha) for estimation of soil moisture (Zhao et al. 2011): The aim of this study was to investigate the potential of a remotely measurable soil evaporation transfer coefficient (ha) for estimation of soil moisture.

$$\mathrm{Mv} = a \times \ln(\mathrm{ha}) + b \tag{2.4}$$

where Mv is volumetric soil moisture, *a* and *b* are regression coefficients, respectively. The soil evaporation transfer coefficient (ha), a coefficient proposed through the equation

$$\mathrm{ha} = \frac{T_\mathrm{s} - T_\mathrm{a}}{T_\mathrm{sd} - T_\mathrm{a}} \tag{2.5}$$

where T_s, T_sd, and T_a are the land surface temperature (LST), dry soil (without evaporation) surface temperature, and air temperature, respectively.

The lessons drawn from the study can be characterized as follows: (i) 'ha' can serve as a useful and easily measurable indicator for evaluating the soil moisture status, and it generally increases as soil moisture decreases, (ii) the method of land surface

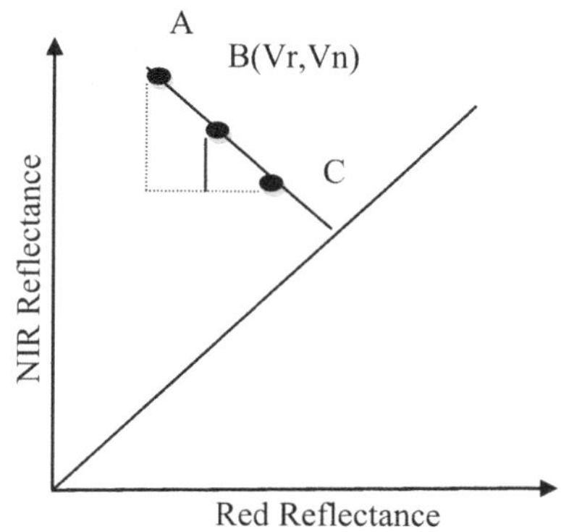

Fig. 2.4 Red-Near infrared feature space

energy balance was superior to the method of maximum temperature in estimating ha. But the latter is simpler in calculation procedures, with reliable accuracy, and (iii) with increasing soil depth, the accuracy of 'ha' for soil moisture estimation dropped due to MODIS's increasing inability to measure it at deeper soil layers.

Linear decomposition of mixture pixels (Gao et al. 2013): This study suggested a method of estimating soil moisture based on the linear decomposition of mixture pixels. In this study, mixture pixel reflectance was considered as a linear combination of vegetation and soil reflectance.

Figure 2.4 shows Red-NIR feature space, where Vr and Vn are the vegetation canopy Red and NIR reflectance, respectively, Sr and Sn are the direct soil reflectance at the Red and NIR band, respectively; Rr and Rn are the reflectance of mixture pixels at the red and NIR band, respectively. It is concluded that the precision of surface soil moisture calculated by the method proposed in this study can satisfy application needs. However, there were some limitations to the soil moisture inversion model in this study. The first is that the minimum and maximum NDVI values were determined by empirical methods which lead to dependency on regional characteristics. The second limitation is that mixture pixel reflectance is considered as the linear combination of vegetation and soil reflectance; however, it is a combination of several factors, and third is that the determination of soil line equation is on the assumption that all soil pixels are located on the constant soil line, with the difference of soil reflectance only caused by soil moisture. But in reality, the soil red and near infrared band reflectance are also influenced by soil texture, the surface roughness (Gao et al. 2013).

The backscatter measured by radar and the emission measured by a radiometer are both very sensitive to the moisture content of bare soil surfaces due to large contrast between dielectric properties of water and soil (Ulaby et al. 1979, 1983; Dobson et al. 1985; Hallikainen et al. 1985; Singh et al. 1990; Krishan et al. 2009; Mishra et al. 2014; Reddy et al. 2017; Watham et al. 2017). In addition, microwave spectrum has soil penetration capability. In active microwave remote sensing, the radar backscattered signal from a bare soil depends on combination of factors, including radar properties (frequency, polarization), incident angle and most importantly the composition and degree of heterogeneity of the surface (Ulaby et al. 1979; Dobson and Ulaby 1981; Du et al. 2000; Hsieh 2001; Verhoest et al. 2008; Wang and Qu 2009; Fang et al. 2013; Fayne et al. 2017; Hashemi et al. 2017). A lot of research has been concentrated in this area during the last two decades (Wigneron et al. 2003; Loew 2008; Hong et al. 2009, 2011; Mladenova et al. 2010; Lakshmi 2013; Parinussa et al. 2016a, b; Kumar et al. 2017; Sutton and Lakshmi 2017). Most of the Earth's materials dielectric constant ranges from 1–3, and water has relatively very high dielectric constant of value 80. Dielectric constant is an indication of reflectivity and conductivity. This is the basic behind measurement of soil moisture through passive microwave remote sensing.

On the relatively large spatial scale, topographical and soil roughness effects are generally found to be small (Gens 2008; Wirth et al. 2012). Therefore in passive remote sensing approach, key variables are vegetation and surface temperature. Different retrieval algorithms using vegetation cover classification (Jackson 1993) two-parameter retrievals, Dual polarization observations (Owe et al. 2001; Meesters et al. 2005; Zhao et al. 2011) have been used for soil moisture retrieval in passive microwave remote sensing. Most soil moisture retrieval approaches to relate radiometer brightness temperature to soil moisture contents are either site-specific or mainly valid for weakly vegetated regions and relatively flat surface. The uncertainty in the relationship between the brightness, temperature, and soil moisture limits the accuracy (Wang and Qu 2009). In active microwave remote sensing, the radar backscattered signal from a bare soil depends on combination of factors, including radar properties (frequency, polarization), surface characteristics (dielectric constant of the soil, and by consequence soil moisture, and surface roughness), and incoming angle of microwave (Verhoest et al. 2008). Furthermore, if the soil is vegetated, additional information about biomass, canopy structure is needed. Table 2.3 presents summary of advantages and limitation of remote sensing data for near-surface soil moisture estimation.

Vegetation over surface and soil roughness has a significant effect on the retrieval methods, which are used to correct soil roughness and vegetation effects (Wigneron et al. 2003). The characterization of soil roughness is not fully understood, and a large range of roughness parameter values can be obtained for the same surface when different measurement methodologies are used (Verhoest et al. 2008). Different empirical approaches as Topp's Universal Calibration and Ledieu's Calibration, semi-empirical approaches as Dielectric-Mixing Model have been developed to establish relations between dielectric constant and surface soil moisture. These are complex, site-specific, or suited for bare soil surface conditions rather than the vegetated surface. The measurements obtained are averages, and there is inconsistency across pixels. The underlying heterogeneity observed at the land surface is therefore not detected (Weihermüller et al. 2007).

With the advancement of classification and regression methods, different machine learning algorithms have been also used for Digital Soil Moisture Mapping (DSMM). The range of methods such as, viz., CDT, SVM, RVM, ANN, Deep learning, KNN algorithms have been adopted by different scholars around the world.

SVM: Method was proposed by Vapnik which rely on the statistical learning theory (Vapnik 1998; 2000), it can effectively avoid overfitting and multidimensional problem. Good performances have been reported in many SVM applications (Yang and Huang 2002; Liu et al. 2008, 2010; Ahmad et al. 2010; Deng et al. 2011; Rodriguez-galiano et al. 2012; Zaman et al. 2012; Srivastava et al. 2013). To estimate soil moisture content using remote sensing data for the selected sites in the Lower Colorado River Basin, Ahmad et al. (2010) developed kernel-based SVM model. Remote sensing data that includes backscatter and incidence angle from Tropical Rainfall Measuring Mission (TRMM) and NDVI from Advanced Very High Resolution Radiometer (AVHRR) are used to estimate soil water content. Both, VIC simulated and observed soil moisture are used to train and test the SVM model. VIC is a macro-scale water and energy balance model that uses meteorological, soil, and vegetation data to estimate gridded surface and subsurface runoff (Liang et al. 1994). The study suggested that SVM method can be used as an alternative to the computationally expensive and data-intensive physical models (Ahmad et al. 2010). Yang and Huang (2002) combine the advantages of time series model and SVM algorithm to explore the applicability of time series-based SVM model for prediction of soil moisture. In this work, trend component and a periodical component of soil moisture were set as the inputs of SVM. Soil moisture data at two depths (0–20 cm, 20–60 cm) was collected between the years 1999 and 2002 in different months for training and testing purposes. Results were also compared with physical model WHNCS. WHNCS is based on the physical models including crop evaporation model, soil water solution model, and soil temperature estimate model (Shirong 2002). For the prediction of soil moisture, the statistical result (t-test) indicates that there is no obvious difference between predicted and observed values in 0–20 cm and 20–60 cm soil layers, and that time series-based SVM model is capable of soil moisture prediction (Yang and Huang 2002). Zaman et al. (2012) applied SVM with Relevance Vector Machines (RVM) to retrieve soil moisture content at two depths (0–6 cm, 0–30 cm). RVM are also known for their robustness, efficiency, and sparseness and provide a statistically sound approach to solve inverse problems and thus to build statistical models. This study combines remotely sensed inputs with field measurements, crop physiological characteristics, soil temperature, soil water-holding capacity, and meteorological data to build a two-step model to estimate soil moisture using both techniques, i.e. RVM and SVM. First, the RVM, as well as SVM, were used to build a model that retrieves surface (0–6 cm) soil moisture content. This information serves as a boundary condition for the second step of this model, which estimates at a depth of 30 cm. The results from the RVM and SVM models were compared and statistics showed that RVM performs better (RMSE = 0.014 m^3/m^3) when compared with SVM (RMSE = 0.017 m^3/m^3) with reduced computational complexity and more suitable real-time implementation (Zaman et al. 2012). Diurnal change in soil temperature and the atmospheric temperature was an important parameter in soil moisture classification using SVM.

ANN: Liu et al. (2008) used ANN model (Kohonen network) in the first layer to classify the input sample into different feature spaces. Then in the second stage,

SVMs combined with different kernel functions were applied to regressing estimate. Due to the use of the multi-stage structure, the proposed hierarchical architecture does not need a large number of the training set. As a result, the computation time for a prediction can be reduced and each prediction can be associated with reliability or confidence of the prediction (Liu et al. 2008).

In one of the study ERS-2 SAR data with incidence angle of radar beam, land cover, surface roughness height, terrain height, leaf area index, and plant water were used to train the feedforward–backpropagation ANN for establishing the relationship between surface soil moisture and terrain as well as sensor variables (Said et al. 2008). The study suggested better correlation for the model developed using backscattering coefficient rather than Digital Number (DN) and in this study leaf area index has been found to be as good as the single bulk variable representing the vegetation characteristics.

ANN is also tested for its ability to predict soil moisture at 1 km resolution. Chai et al. (2010) utilized information on the variability of soil moisture, in terms of its mean and standard deviation to improve the current retrieval accuracy of the ANN method.

Microwave Radiation Imager (MWRI) instrument, on board the FY-3 (Fengyun-3B) series of CMA/NSMC (China Meteorological Administration/National Satellite Meteorological Center) provides global observations of land surface soil moisture. The soil moisture products over Tibetan Plateau were validated using in situ soil moisture measurement networks and Essential Climate Variable (ECV) soil moisture product (Cui et al. 2016). FY-3B/MWRI and ECV product showed almost the same correlation with the in situ measurements. However, the ascending FY-3B/MWRI product has better performance than the ECV product in under the lower NDVI condition (Cui et al. 2016). Uncertainty in the ascending FY-3B/MWRI product increases with increasing NDVI. In Cui et al. (2016) study, FY-3B/MWRI soil moisture product is also reconstructed using the BackPropagation neural Network (BP-NN) based on reconstructed MODIS products, i.e. LST, NDVI, and albedo.

KNN: Coopersmith et al. (2016) generate models of 5 cm soil moisture, with 10 cm soil moisture measurements and antecedent precipitation as inputs, via KNN machine learning techniques. United States Climate Reference Network (USCRN) was used to collect data of 10 cm soil moisture (measured in a cubic meter of water per cubic meter of soil) and precipitation. Validation was conducted with the available, in situ, 5 cm resources and good Root-Mean-Squared-Error (RMSE) of 0.0215 m^3/m^3, was achieved. By demonstrating the capacity to generate accurate 5 cm estimates from antecedent local precipitation and 10 cm sensors located below, a potential replacement methodology emerges. This approach can help ensure robust continuous data records and also an emphasis on soil moisture prediction modelling using machine learning (Coopersmith et al. 2016).

Deep Learning: In recent years, as a new technique of machine learning, deep learning has won numerous contests. The deep learning model has the advantage of representing nonlinearities and modelling complex relationships from large-scale data. To Song et al. (2016) used for surface soil moisture modelling. This study proposed a

DBN (Deep Belief Network)-based MCA (Macroscopic Cellular Automata) model to predict the spatiotemporal patterns of SMC over an irrigated cornfield and results were compared with another hybrid model, Macroscopic Cellular Automata-Multi Layer Perceptron (MCA-MLP). DBNs were employed to achieve the transition functions depending on climate, topography, soil properties, and evaporation. The authors showed that the DBN-MCA model performs better than the MLP-MCA model, and the DBN-MCA model provides a powerful tool for predicting SMC in highly non-linear forms. Moreover, because modelling soil moisture by using environmental variables is gaining increasing popularity, DBN techniques could contribute a lot to enhancing the calibration of MCA-based SMC estimations and hence provide an alternative approach for SMC monitoring in irrigation systems on the basis of canals.

In a recent study, deep learning was also used for upscaling of surface soil moisture (Zhang et al. 2017). Based on abundant satellite remote sensing and ground measurements and the compelling advantages of the deep learning techniques, we upscaled soil measurements and the compelling advantages of the deep learning techniques, Zhang et al. (2017) upscaled soil moisture ground measurements using Visible Infrared Imaging Radiometer Suite (VIIRS) Raw Data Records (RDR) to achieve a national surface soil moisture product at 750 m resolution. Results demonstrate that the estimated models captured the complex relationship between the remote sensing variables and in situ surface soil moisture with RMSE of 0.0084. The results were more accurate than the Soil Moisture Active Passive (SMAP) active radar soil moisture products and the Global Land data assimilation system (GLDAS) 0–10 cm depth soil moisture data (Zhang et al. 2017).

References

Ahmad, S., Kalra, A., & Stephen, H. (2010). Estimating soil moisture using remote sensing data: A machine learning approach. *Advances in Water Resources, 33*(1), 69–80.

An, Q., & Yang, B. (2007). A multicrop identification model based on stepwise removal learning-support vector machine using remote sensing images. *New Zealand Journal of Agricultural Research, 50*(5), 1013–1019.

Andrade, O., Kappas, M., & Erasmi, S. (2010). Assessment of erosion hazard in Torres municipality of Laras tate (Venezuela) based on GIS. *Interciencia, 35*(5), 348–356.

Bagheri Bodaghabadi, M., Martínez-Casasnovas, J., Salehi, M. H., Mohammadi, J., Esfandiarpoor Borujeni, I., Toomanian, N., et al. (2015). Digital soil mapping using artificial neural networks and terrain related attributes. *Pedosphere, 25*(4), 580–591.

Balkovič, J., Čemanová, G., & Kollár, J. (2007). Mapping soils using the fuzzy approach and regression-kriging case study from the Považský Inovec Mountains, Slovakia. *Soil and Water Research, 2007*(4), 123–134.

Bansal, S., Srivastav, S. K., Roy, P. S., & Krishnamuthy, Y. V. N. (2016). An analysis of land use and land cover dynamics and causative drivers in a thickly populated Yamuna river basin of India. *Applied Ecology and Environmental Research, 14*(3), 773–792.

Bauböck, R., Karpenstein-Machan, M., and Kappas, M., 2014. Computing the biomass potentials for maize and two alternative energy crops, triticale and cup plant (Silphium perfoliatum L.), with the crop model BioSTAR in the region of Hannover (Germany). *Environmental Sciences Europe,* 26(1), 1–12.

Beg, M. K., Srivastav, S. K., Carranza, E. J. M., & de Smeth, J. B. (2011). High fluoride incidence in groundwater and its potential health effects in parts of Raigarh District, Chhattisgarh, India. *Current Science, 100*(5), 750–754.

Bell, J. C., Grigal, D. F., & Bates, P. C. (2000). A soil terrain model for estimating spatial patterns of soil organic carbon. In I. Gallant (Ed.), *Terrain analysis-principles and applications* (pp. 295–310). John Wiley & Sons, New York.

Bhakar, R., Srivastav, S. K., Garg, R. D., & Jetten, V. G. (2012). Upscaling soil-hydrologic parameters in sandy desert landscape—An input for distributed hydrological modelling. *Asian Journal of Geoinformatics, 12*(1), 1–10.

Bishop, T. F. A., & McBratney, A. B. (2001). A comparison of prediction methods for the creation of field-extent soil property maps. *Geoderma, 103*(1–2), 149–160.

Blahwar, B., Srivastav, S. K., & de Smeth, J. B. (2012). Use of high-resolution satellite imagery for investigating acid mine drainage from artisanal coal mining in North-Eastern India. *Geocarto International, 27*(3), 231–247.

Boerner, R. E. J., Morris, S. J., Sutherland, E. K., & Hutchinson, T. F. (2000). Spatial variability in soil nitrogen dynamics after prescribed burning in Ohio mixed-oak forests. *Landscape Ecology, 15*(5), 425–439.

Boloorani, A. D., Erasmi, S., & Kappas, M. (2008). Multi-source remotely sensed data combination: Projection transformation gap-fill procedure. *Sensors, 8*(7), 4429–4440.

Bourennane, H., King, D., & Couturier, A. (2000). Comparison of kriging with external drift and simple linear regression for predicting soil horizon thickness with different sample densities. *Geoderma, 97*(3–4), 255–271.

Breiman, L. (2001). Random forest. *Machine Learning, 45*(1), 5–32.

Breiman, L., Friedman, J. H., Olshen, R. A., & Stone, C. J. (1984). *Classification and regression trees*. Cole Publishing, Monterey. California, USA: Wadsworth and Brooks/Cole.

Bui, E. N., & Moran, C. J. (2001). Disaggregation of polygons of surficial geology and soil maps using spatial modelling and legacy data. *Geoderma, 103*(1–2), 79–94.

Bui, E. N., & Moran, C. J. (2003). A strategy to fill gaps in soil survey over large spatial extents: An example from the Murray-Darling Basin of Australia. *Geoderma, 111*(1–2), 21–44.

Bui, E. N., Loughhead, A., & Corner, R. (1999). Extracting soil-landscape rules from previous soil surveys. *Australian Journal of Soil Research, 37*(3), 495.

Burrough, P. A., Van Gaans, P. F. M., & Hootsmans, R. (1997). Continuous classification in soil survey: Spatial correlation, confusion and boundaries. *Geoderma, 77*(2–4), 115–135.

Campling, P., Gobin, A., & Feyen, J. (2002). Logistic modeling to spatially predict the probability of soil drainage classes. *Soil Science Society of America Journal, 66*(4), 1390–1401.

Carré, F., & Girard, M. C. (2002). Quantitative mapping of soil types based on regression kriging of taxonomic distances with landform and land cover attributes. *Geoderma, 110*(3–4), 241–263.

Carre, F., McBratney, A. B., Mayr, T., & Montanarella, L. (2007). Digital soil assessments: Beyond DSM. *Geoderma, 142*(1–2), 69–79.

Chai, S.-S., Walker, J., Makarynskyy, O., Kuhn, M., Veenendaal, B., & West, G. (2010). Use of soil moisture variability in artificial neural network retrieval of soil moisture. *Remote Sensing, 2*(1), 166–190.

Chanasyk, D. S., & Naeth, M. A. (1996). Field measurement of soil moisture using neutron probes. *Canadian Journal of Soil Science, 76*(3), 317–323.

Chaplot, V., Walter, C., & Curmi, P. (2000). Improving soil hydromorphy prediction according to DEM resolution and available pedological data. *Geoderma, 97*(3–4), 405–422.

Chaudhary, S. K., Kumar, D., & Jain, M. K. (2016a). Multi-classifier fusion for land use land cover mapping in Jharia Coal Field. In *Geostatistical and geospatial approaches for the characterization of natural resources in the environment* (pp. 773–777). Cham: Springer International Publishing.

Chaudhary, S. K., Kumar, D., & Jain, M. K. (2016b). Performance analysis of hyperspherical colour sharpening method for IRS satellite images. *Imaging Science Journal, 64*(6), 305–312.

Chen, J., Wan, S., Henebry, G., Qi, J., Sun, G., Kappas, M., et al. (2013). *Dryland East Asia: Land dynamics amid social and climate change*. Berlin, Boston: De Gruyter: Together with Higher Education Press.

ChenChi, F., ChenCheng, R., & Son Nguyen, T. (2012). Investigating rice cropping practices and growing areas from MODIS data using empirical mode decomposition and support vector machines. *GIScience & Remote Sensing, 49*(1), 117–138.

Cialella, A., Dubayah, R., Lawrence, W., & Levine, E. (1997). Predicting soil drainage class using remotely sensed and digital elevation data. *Photogrammetric Engineering and Remote Sensing, 63*(2), 171–177.

Connolly, J., & Holden, N. M. (2009). Mapping peat soils in Ireland: updating the derived Irish peat map. *Irish Geography, 42*(3), 343–352.

Coopersmith, E. J., Cosh, M. H., Bell, J. E., & Boyles, R. (2016). Using machine learning to produce near surface soil moisture estimates from deeper in situ records at U.S. Climate Reference Network (USCRN) locations: Analysis and applications to AMSR-E satellite validation. *Advances in Water Resources, 98*, 122–131.

Cui, Y., Long, D., Hong, Y., Zeng, C., Zhou, J., Han, Z., et al. (2016). Validation and reconstruction of FY-3B/MWRI soil moisture using an artificial neural network based on reconstructed MODIS optical products over the Tibetan Plateau. *Journal of Hydrology, 543*, 242–254.

Dashora, A., Sreenivas, B., Lohani, B., Malik, J. N., & Shah, A. A. (2006). GCP collection for CORONA satellite photographs: Issues and methodology. *Journal of the Indian Society of Remote Sensing, 34*(2), 153–160.

Dashora, A., Lohani, B., & Deb, K. (2013). Two-step procedure of optimisation for flight planning problem for airborne LiDAR data acquisition. *International Journal of Mathematical Modelling and Numerical Optimisation, 4*(4), 323.

Deng, J., Chen, X., Du, Z., & Zhang, Y. (2011). Soil water simulation and predication using stochastic models based on LS-SVM for red soil region of China. *Water Resources Management, 25*(11), 2823–2836.

Deschamps, B., McNairn, H., Shang, J., & Jiao, X. (2012). Towards operational radar-only crop type classification: Comparison of a traditional decision tree with a random forest classifier. *Canadian Journal of Remote Sensing, 38*(1), 60–68.

Dietterich, T. G. (2000). An experimental comparison of three methods for constructing ensembles of decision trees. *Machine Learning, 40*, 139–157.

Dobos, E., Micheli, E., Baumgardner, M. F., Biehl, L., & Helt, T. (2000). Use of combined digital elevation model and satellite radiometric data for regional soil mapping. *Geoderma, 97*(3–4), 367–391.

Dobriyal, P., Qureshi, A., Badola, R., & Hussain, S. A. (2012). A review of the methods available for estimating soil moisture and its implications for water resource management. *Journal of Hydrology, 458–459*, 110–117.

Dobson, M. C., & Ulaby, F. T. (1981). Microwave backscatter dependence on surface roughness, soil moisture and soil texture: Part III—Soil tension. *IEEE Transactions on Geoscience and Remote Sensing, 19*(1), 51–61.

Dobson, M. C., Ulaby, F. T., Hallikainen, M. T., & El-Rayes, M. A. (1985). Microwave dielectric behavior of wet soil-part II: Dielectric mixing models. *IEEE Transactions on Geoscience and Remote Sensing, GE-23*(1), 35–46.

Du, Y., Ulaby, F. T., & Dobson, M. C. (2000). Sensitivity to soil moisture by active and passive microwave sensors. *IEEE Transactions on Geoscience and Remote Sensing, 38*(1), 105–114.

Dukes, M. D., Zotarelli, L., & Morgan, K. T. (2010). Use of irrigation technologies for vegetable crops in Florida. *Horttechnology, 20*(1), 133–142.

Erasmi, S., Twele, A., Ardiansyah, M., Malik, A., & Kappas, M. (2004). Mapping deforestation and land cover conversion at the rainforest margin in central Sulawesi, Indonesia. *EARSeL eProceedings, 3*(3), 388–397.

Erlingsson, S., Baltzer, S., Baena, J., & Bjarnason, G. (2009). Measurement techniques for water flow. In *Water in road structures* (pp. 45–67). Dordrecht: Springer.

Fang, B., Lakshmi, V., Bindlish, R., Jackson, T. J., Cosh, M., & Basara, J. (2013). Passive microwave soil moisture downscaling using vegetation index and skin surface temperature. *Vadose Zone Journal, 12*(3), 1–19.

Fayne, J. V., Bolten, J. D., Doyle, C. S., Fuhrmann, S., Rice, M. T., Houser, P. R., et al. (2017). Flood mapping in the lower Mekong River Basin using daily MODIS observations. *International Journal of Remote Sensing, 38*(6), 1737–1757.

Finke, P. A. (2012). On digital soil assessment with models and the Pedometrics agenda. *Geoderma, 171–172*, 3–15.

Gao, Z., Xu, X., Wang, J., Yang, H., Huang, W., & Feng, H. (2013). A method of estimating soil moisture based on the linear decomposition of mixture pixels. *Mathematical and Computer Modelling, 58*(3–4), 606–613.

Gens, R. (2000). The influence of input parameters on SAR interferometric processing and its implication on the calibration of SAR interferometric data. *International Journal of Remote Sensing, 21*(8), 1767–1771.
Gens, R. (2003). Two-dimensional phase unwrapping for radar interferometry: Developments and new challenges. *International Journal of Remote Sensing, 24*(4), 703–710.
Gens, R. (2008). Oceanographic applications of SAR remote sensing. *GIScience & Remote Sensing, 45*(3), 275–305.
Gessler, P. E., Moore, I. D., Mckenzie, N. J., & Ryan, P. J. (1995). Soil-landscape modelling and spatial prediction of soil attributes. *International Journal of Geographical Information Systems, 9*(4), 421–432.
Ghosh, S., & Lohani, B. (2013). Mining lidar data with spatial clustering algorithms. *International Journal of Remote Sensing, 34*(14), 5119–5135.
Giasson, E., Clarke, R. T., Inda Junior, A. V., Merten, G. H., & Tornquist, C. G. (2006). Digital soil mapping using multiple logistic regression on terrain parameters in southern Brazil. *Scientia Agricola, 63*(3), 262–268.
Giraldo, M. A., Bosch, D., Madden, M., Usery, L., & Kvien, C. (2008). Landscape complexity and soil moisture variation in south Georgia, USA, for remote sensing applications. *Journal of Hydrology, 357*(3–4), 405–420.
Grimm, R., Behrens, T., Märker, M., & Elsenbeer, H. (2008). Soil organic carbon concentrations and stocks on Barro Colorado Island—Digital soil mapping using random forests analysis. *Geoderma, 146*(1–2), 102–113.
Hallikainen, M. T., Ulaby, F. T., Dobson, M. C., El-Rayes, M. A., & Wu, L.-K. (1985). Microwave dielectric behavior of wet soil-part I: Empirical models and experimental observations. *IEEE Transactions on Geoscience and Remote Sensing, GE-23*(1), 25–34.
Hashemi, H., Nordin, M., Lakshmi, V., Huffman, G. J., & Knight, R. (2017). Bias correction of long-term satellite monthly precipitation product (TRMM 3B43) over the conterminous United States. *Journal of Hydrometeorology, 18*(9), 2491–2509.
Hengl, T., Toomanian, N., Reuter, H. I., & Malakouti, M. J. (2007). Methods to interpolate soil categorical variables from profile observations: Lessons from Iran. *Geoderma, 140*(4), 417–427.
Hong, S., Lakshmi, V., Small, E. E., Chen, F., Tewari, M., & Manning, K. W. (2009). Effects of vegetation and soil moisture on the simulated land surface processes from the coupled WRF/Noah model. *Journal of Geophysical Research, 114*(D18), D18118.
Hong, S., Lakshmi, V., Small, E. E., & Chen, F. (2011). The influence of the land surface on hydrometeorology and ecology : new advances from modeling and satellite remote sensing.*Hydrology Research, 42.2*(3), 95–113.
Hsieh, C.-Y. (2001). Microwave backscattering model for a bare soil field. *Electromagnetics, 21*(3), 259–273.
Hulley, G. C., Hook, S. J., & Baldridge, A. M. (2010). Investigating the effects of soil moisture on thermal infrared land surface temperature and emissivity using satellite retrievals and laboratory measurements. *Remote Sensing of Environment, 114*(7), 1480–1493.
Hutchinson, T. F., Boerner, R. E. J., Iverson, L. R., Sutherland, S., & Sutherland, E. K. (1999). Landscape patterns of understory composition and richness across a moisture and nitrogen mineralization gradient in Ohio (U.S.A.) Quercus forests. *Plant Ecology, 144*(2), 177–189.
Iverson, L. R., Dale, M. E., Scott, C. T., & Prasad, A. (1997). A GIS-derived integrated moisture index to predict forest composition and productivity of Ohio forests (U.S.A.). *Landscape Ecology, 12*(5), 331–348.
Jackson, T. J. (1993). Measuring surface soil moisture using passive microwave remote sensing. *Hydrological Processes, 7*(2), 139–152.
Jafari, A., Ayoubi, S., Khademi, H., Finke, P. A., & Toomanian, N. (2013). Selection of a taxonomic level for soil mapping using diversity and map purity indices: A case study from an Iranian arid region. *Geomorphology, 201*, 86–97.
Jafari, A., Finke, P. A., Vande Wauw, J., Ayoubi, S., & Khademi, H. (2012). Spatial prediction of USDA-great soil groups in the arid Zarand region, Iran: Comparing logistic regression approaches to predict diagnostic horizons and soil types. *European Journal of Soil Science, 63*(2), 284–298.
Jenny, H. (1941). *Factors of soil formation. A system of quantitative pedology*. McGraw-Hill Book Company. New York.

Jeyaseelan, A. T., Roy, P. S., & Young, S. S. (2007). Persistent changes in NDVI between 1982 and 2003 over India using AVHRR GIMMS (Global Inventory Modeling and Mapping Studies) data. *International Journal of Remote Sensing, 28*(21), 4927–4946.

Kale, M. P., Chavan, M., Pardeshi, S., Joshi, C., Verma, P. A., Roy, P. S., et al. (2016). Land-use and land-cover change in Western Ghats of India. *Environmental Monitoring and Assessment, 188*(7).

Kappas, M. (2013). Estimation of global bioenergy potentials and their contribution to the world's future energy demand—a short review. In H. Ruppert, M. Kappas, & J. Ibendorf (Eds.), *Sustainable bioenergy production—An integrated approach* (pp. 75–95). Dordrecht: Springer Netherlands.

Kempen, B., Brus, D. J., & Heuvelink, G. B. M. (2012). Soil type mapping using the generalised linear geostatistical model: A case study in a Dutch cultivated peatland. *Geoderma, 189–190*, 540–553.

Khanna, S., Palacios-Orueta, A., Whiting, M. L., Ustin, S. L., Riaño, D., & Litago, J. (2007). Development of angle indexes for soil moisture estimation, dry matter detection and land-cover discrimination. *Remote Sensing of Environment, 109*(2), 154–165.

Kim, J., Grunwald, S., Rivero, R. G., & Robbins, R. (2012). Multi-scale modeling of soil series using remote sensing in a wetland ecosystem. *Soil Science Society of America Journal, 76*(6), 2327.

Korolyuk, T. V., & Shcherbenko, H. V. (2007). Compiling soil maps on the basis of remotely-sensed data digital processing: soil interpretation. *International Journal of Remote Sensing, 15*(7), 1379–1400.

Kothapalli Venkata, R., Poloju, S., Mullapudi Venkata Rama, S. S., Gogineni, A., Prabir Kumar, D., Allakki Venkata, R., et al. (2017). Multi-incidence angle RISAT-1 Hybrid Polarimetric SAR data for large area mapping of maize crop—A case study in Khagaria district, Bihar, India. *International Journal of Remote Sensing, 38*(20), 5487–5501.

Kovačević, M., Bajat, B., & Gajić, B. (2010). Soil type classification and estimation of soil properties using support vector machines. *Geoderma, 154*(3–4), 340–347.

Kravchenko, A. N. (2008). Mapping of soil drainage classes using topographical data and soil electrical conductivity. *Handbook of Agricultural Geophysics* (1), 255–261.

Krishan, G., Srivastav, S. K., Kumar, S., Saha, S. K., & Dadhwal, V. K. (2009). Quantifying the underestimation of soil organic carbon by the Walkley and Black technique—Examples from Himalayan and Central Indian soils. *Current Science, 96*(8), 1133–1136.

Kumar, D. (2015a). Fair allocation of multi-resources for multi-class users in cloud computing. In *International Conference on Computing, Communication and Automation (ICCCA2015)* (pp. 661–663). Greater Noida, India: IEEE.

A survey on resource allocation techniques in cloud computing. In *International Conference on Computing, Communication and Automation (ICCCA2015)* (pp. 655–660). Greater Noida, India: IEEE.

Kumar, D., Deb, D., & Mamgain, R. (2015a). Analysis of different multiprocessor architectures for radar signal processing with performance metrics. In *IEEE Radar Conference* (pp. 289–294). Johannesburg, South Africa: IEEE.

Kumar, P., Gupta, D. K., Mishra, V. N., & Prasad, R. (2015b). Comparison of support vector machine, artificial neural network, and spectral angle mapper algorithms for crop classification using LISS IV data. *International Journal of Remote Sensing, 36*(6), 1604–1617.

Kumar, P., Prasad, R., Choudhary, A., Mishra, V. N., Gupta, D. K., & Srivastava, P. K. (2016). A statistical significance of differences in classification accuracy of crop types using different classification algorithms. *Geocarto International, 6049*, 1–19.

Kumar, B., Patra, K. C., & Lakshmi, V. (2017). Error in digital network and basin area delineation using D8 method: A case study in a sub-basin of the Ganga. *Journal of Geological Society of India, 89*, 65–70.

Lagacherie, P., Legros, J. P., & Burfough, P. A. (1995). A soil survey procedure using the knowledge of soil pattern established on a previously mapped reference area. *Geoderma, 65*(3–4), 283–301.

Lakshmi, V. (2013). Remote sensing of soil moisture. *ISRN Soil Science*, 1–33.

Liang, X., Lettenmaier, D. P., Wood, E. F., & Burges, S. J. (1994). A simple hydrologically based model of land surface water and energy fluxes for general circulation models. *Journal of Geophysical Research, 99*(D7), 14415.

Liu, H., Xie, D., & Wu, W. (2008). Soil water content forecasting by ANN and SVM hybrid architecture. *Environmental Monitoring and Assessment, 143*(1–3), 187–193.

Liu, D., Yu, Z., & Lü, H. (2010). Data assimilation using support vector machines and ensemble Kalman filter for multi-layer soil moisture prediction. *Water Science and Engineering, 3*(4), 361–377.

Loew, A. (2008). Impact of surface heterogeneity on surface soil moisture retrievals from passive microwave data at the regional scale: The Upper Danube case. *Remote Sensing of Environment, 112*(1), 231–248.

Lohani, B., & Singh, R. (2008). Effect of data density, scan angle, and flying height on the accuracy of building extraction using LiDAR data. *Geocarto International, 23*(2), 81–94.

Lohani, B., Mason, D. C., Scott, T. R., & Sreenivas, B. (2006). Extraction of tidal channel networks from aerial photographs alone and combined with laser altimetry. *International Journal of Remote Sensing, 27*(1), 5–25.

Lunt, I. A., Hubbard, S. S., & Rubin, Y. (2005). Soil moisture content estimation using ground-penetrating radar reflection data. *Journal of Hydrology, 307*(1–4), 254–269.

Mallick, K., Bhattacharya, B. K., & Patel, N. K. (2009). Estimating volumetric surface moisture content for cropped soils using a soil wetness index based on surface temperature and NDVI. *Agricultural and Forest Meteorology, 149*(8), 1327–1342.

Marchetti, A., Piccini, C., Santucci, S., Chiuchiarelli, I., & Francaviglia, R. (2011). Simulation of soil types in Teramo province (Central Italy) with terrain parameters and remote sensing data. *CATENA, 85*(3), 267–273.

Mathur, A., & Foody, G. M. (2008). Crop classification by support vector machine with intelligently selected training data for an operational application. *International Journal of Remote Sensing, 29*(8), 2227–2240.

McBratney, A. B., Mendonça Santos, M. L., & Minasny, B. (2003). On digital soil mapping. *Geoderma, 117*(1–2), 3–52.

McKenzie, N. J., & Ryan, P. J. (1999). Spatial prediction of soil properties using environmental correlation. *Geoderma, 89*(1–2), 67–94.

Meesters, A. G. C. A., DeJeu, R. A. M., & Owe, M. (2005). Analytical derivation of the vegetation optical depth from the microwave polarization difference index. *IEEE Geoscience and Remote Sensing Letters, 2*(2), 121–123.

Mishra, M. D., Patel, P., Srivastava, H. S., Patel, P. R., Shukla, A., & Shukla, A. K. (2014). Absolute radiometric calibration of FRS-1 and MRS mode of RISAT-1 synthetic aperture radar (SAR) data using corner reflectors. *International Journal of Advanced Engineering Research and Science, 1*(6), 78–89.

Mladenova, I., Lakshmi, V., Walker, J. P., Panciera, R., Wagner, W., & Doubkova, M. (2010). Validation of the ASAR global monitoring mode soil moisture product using the NAFE ' 05 data set. *IEEE Transactions on Geoscience and Remote Sensing, 48*(6), 2498–2508.

Mondal, M. S., Sharma, N., Garg, P. K., & Kappas, M. (2016). Statistical independence test and validation of CA Markov land use land cover (LULC) prediction results. *Egyptian Journal of Remote Sensing and Space Science, 19*(2), 259–272.

Moonjun, R., Farshad, A., Shrestha, D. P., & Vaiphasa, C. (2010). Artificial neural network and decision tree in predictive soil mapping of Hoi Num Rin sub-watershed, Thailand. In *Digital soil mapping* (pp. 151–164). Dordrecht: Springer Netherlands.

Moran, C. J., & Bui, E. N. (2002). Spatial data mining for enhanced soil map modelling. *International Journal of Geographical Information Science, 16*(6), 533–549.

Morris, S. J., & Boerner, R. E. J. (1998). Landscape patterns of nitrogen mineralization and nitrification in southern Ohio hardwood forests. *Landscape Ecology, 13*(4), 215–224.

Muñoz-Carpena, R. (2015). Field devices for monitoring soil water content [online]. *EDIS Publication BUL343*. Retrieved November 21, 2015 from https://edis.ifas.ufl.edu/ae266.

Murthy, C. S., Raju, P. V., & Badrinath, K. V. S. (2003). Classification of wheat crop with multi-temporal images: Performance of maximum likelihood and artificial neural networks. *International Journal of Remote Sensing, 24*(23), 4871–4890.

Noi, P., Kappas, M., & Degener, J. (2016). Estimating daily maximum and minimum land air surface temperature using MODIS land surface temperature data and ground truth data in Northern Vietnam. *Remote Sensing, 8*(12), 1002.

Odeh, I. O. A., & McBratney, A. B. (2000). Using AVHRR images for spatial prediction of clay content in the lower Namoi Valley of eastern Australia. *Geoderma, 97*(3–4), 237–254.

Ohlmacher, G. C., & Davis, J. C. (2003). Using multiple logistic regression and GIS technology to predict landslide hazard in northeast Kansas, USA. *Engineering Geology, 69*(3–4), 331–343.

Owe, M., De Jeu, R., & Walker, J. (2001). A methodology for surface soil moisture and vegetation optical depth retrieval using the microwave polarization difference index. *IEEE Transactions on Geoscience and Remote Sensing, 39*(8), 1643–1654.

Pahlavan Rad, M. R., Toomanian, N., Khormali, F., Brungard, C. W., Komaki, C. B., & Bogaert, P. (2014). Updating soil survey maps using random forest and conditioned Latin hypercube sampling in the loess derived soils of northern Iran. *Geoderma, 232–234*, 97–106.

Pal, M. (2008). Ensemble of support vector machines for land cover classification. *International Journal of Remote Sensing, 29*(10), 3043–3049.

Pal, M., & Mather, P. M. (2003a). Support vector classifiers for land cover classification. In *Map India Conference* (pp. 1–11). Aligarh, Uttar Pradesh.

Pal, M., & Mather, P. M. (2003b). An assessment of the effectiveness of decision tree methods for land cover classification. *Remote Sensing of Environment, 86*(4), 554–565.

Pal, M., & Mather, P. M. (2005). Support vector machines for classification in remote sensing. *International Journal of Remote Sensing, 26*(5), 1007–1011.

Palacios-Orueta, A., Khanna, S., & Litago, J. (2005). Assessment of NDVI and NDWI spectral indices using MODIS time series analysis and development of a new spectral index based on MODIS shortwave infrared bands. In *1st International Conference of Remote Sensing and Geoinformation Processing*. Trier, Germany.

Paloscia, S., Pampaloni, P., Pettinato, S., & Santi, E. (2008). A comparison of algorithms for retrieving soil moisture from ENVIS AT/AS AR images. *IEEE Transactions on Geoscience and Remote Sensing, 46*(10), 3274–3284.

Panagos, P., Jones, A., Bosco, C., & Senthil Kumar, P. S. (2011). European digital archive on soil maps (EuDASM): preserving important soil data for public free access. *International Journal of Digital Earth, 4*(5), 434–443.

Pandey, J., Kumar, D., Mishra, R. K., Mohalik, N. K., Khalkho, A., & Singh, V. K. (2013). Application of thermography technique for assessment and monitoring of coal mine fire: A special reference to Jharia Coal Field, Jharkhand, India. *International Journal of Advanced Remote Sensing and GIS, 2*(1), 138–147.

Pandey, J., Mohalik, N. K., Mishra, R. K., Khalkho, A., Kumar, D., & Singh, V. K. (2015). Investigation of the role of fire retardants in preventing spontaneous heating of coal and controlling coal mine fires. *Fire Technology, 51*(2), 227–245.

Pandey, J., Kumar, D., Singh, V. K., & Mohalik, N. K. (2016). Environmental and socio-economic impacts of fire in Jharia coalfield, Jharkhand, India: An appraisal. *Current Science, 110*(9), 19–23.

Parinussa, R., Lakshmi, V., Johnson, F., & Sharma, A. (2016a). Comparing and combining remotely sensed land surface temperature products for improved hydrological applications. *Remote Sensing, 8*(2), 162.

Parinussa, R. M., Lakshmi, V., Johnson, F. M., & Sharma, A. (2016b). A new framework for monitoring flood inundation using readily available satellite data. *Geophysical Research Letters, 43*(6), 2599–2605.

Pásztor, L., Szabó, J., Bakacsi, Z., Matus, J., & Laborczi, A. (2012). Compilation of 1:50,000 scale digital soil maps for Hungary based on the digital Kreybig soil information system. *Journal of Maps, 8*(3), 215–219.

Patel, P., Srivastava, H. S., Panigrahy, S., & Parihar, J. S. (2006). Comparative evaluation of the sensitivity of multi-polarized multi-frequency SAR backscatter to plant density. *International Journal of Remote Sensing, 27*(2), 293–305.

Pires, L. F., Bacchi, O. O. S., & Reichardt, K. (2005). Soil water retention curve determined by gamma-ray beam attenuation. *Soil and Tillage Research, 82*(1), 89–97.

Poggio, L., Gimona, A., & Brewer, M. J. (2013). Regional scale mapping of soil properties and their uncertainty with a large number of satellite-derived covariates. *Geoderma, 209–210*, 1–14.

Prasad, A. M., Iverson, L. R., & Liaw, A. (2006). Newer classification and regression tree techniques: Bagging and random forests for ecological prediction. *Ecosystems, 9*(2), 181–199.

Punera, K., & Ghosh, J. (2008). Consensus-based ensembles of soft clusterings. *Applied Artificial Intelligence, 22*(7–8), 780–810.

Ranjan, V., Sen, P., Kumar, D., & Sarsawat, A. (2015). A review on dump slope stabilization by revegetation with reference to indigenous plant. *Ecological Processes, 4*(1), 14.

Rätsch, G., Onoda, T., & Müller, K. R. (2001). Soft margins for AdaBoost. *Machine Learning, 42*(3), 287–320.

Ravan, S., & Roy, P. S. (1997). Satellite remote sensing for the ecological analysis of peatbog areas in Lower Saxony. *Plant Ecology, 131*, 129–141.

Ravan, S., Roy, P. S., & Sharma, C. M. (1995). Space remote-sensing for spatial vegetation characterization. *Journal of Biosciences, 20*(3), 427–438.

Ravan, S., Kale, M., & Roy, P. S. (2004). Identification of potential sites for in situ conservation of landraces associated with forest ecosystem—Geomatics approach. *Current Science, 87*(8), 1115–1122.

Reddy, C. S. S., Bhattacharya, A., & Srivastav, S. K. (1993a). Night-time TM short wavelength infrared data analysis of Barren Island volcano, South Andaman, India. *International Journal of Remote Sensing, 14*(4), 783–787.

Reddy, C. S. S., Srivastav, S. K., & Bhattacharya, A. (1993b). Application of thematic mapper short wavelength infrared data for the detection and monitoring of high temperature related geoenvironmental features. *International Journal of Remote Sensing, 14*(17), 3125–3132.

Reddy, C. S., Singh, S., Dadhwal, V. K., Jha, C. S., Rao, N. R., & Dwakar, P. G. (2017). Predictive modelling of the spatial pattern of past and future forest cover changes in India. *Journal of Earth System Science, 126*(1), 8.

Richardson, M. D., Meisner, C. A., Hoveland, C. S., & Karnok, K. J. (1992). Time domain reflectometry in closed container studies. *Agronomy Journal, 84*(6), 1061–1063.

Rodriguez-galiano, V. F., Ghimire, B., Rogan, J., Chica-olmo, M., & Rigol-sanchez, J. P. (2012). An assessment of the effectiveness of a random forest classifier for land-cover classification. *ISPRS Journal of Photogrammetry and Remote Sensing, 67*, 93–104.

Rodriguez-Galiano, V. F., & Chica-Rivas, M. (2014). Evaluation of different machine learning methods for land cover mapping of a Mediterranean area using multi-seasonal Landsat images and Digital Terrain Models. *International Journal of Digital Earth, 7*(6), 492–509.

Roy, P. S., & Ravan, S. A. (1996). Biomass estimation using satellite remote sensing data—An investigation on possible approaches for natural forest. *Journal of Biosciences, 21*(4), 535–561.

Said, S., Kothyari, U. C., & Arora, M. K. (2008). ANN-based soil moisture retrieval over bare and vegetated areas using ERS-2 SAR data. *Journal of Hydrologic Engineering, 13*(6), 461–475.

Sarkar, A., Majumdar, A., Chatterjee, S., Chatterjee, D., Ray, S. S., & Kartikeyan, B. (2008). Study of the potential of alternative crops by integration of multisource data using a neuro-fuzzy technique. *International Journal of Remote Sensing, 29*(795405157), 5479–5493.

Schrott, L., & Sass, O. (2008). Application of field geophysics in geomorphology: Advances and limitations exemplified by case studies. *Geomorphology, 93*(1–2), 55–73.

Scull, P., Franklin, J., & Chadwick, O. A. (2005). The application of classification tree analysis to soil type prediction in a desert landscape. *Ecological Modelling, 181*(1), 1–15.

Seneviratne, S. I., Corti, T., Davin, E. L., Hirschi, M., Jaeger, E. B., Lehner, I., et al. (2010). Investigating soil moisture-climate interactions in a changing climate: A review. *Earth-Science Reviews, 99*(3–4), 125–161.

Shirong, Z. (2002). *GIS-based simulation of regional soil water and nitrogen behavior and analysis of agricultural management*. Beijing: China Agriculture University.

Shukla, G., Garg, R. D., Srivastava, H. S., & Garg, P. K. (2018). Performance analysis of different predictive models for crop classification across an aridic to ustic area of Indian states. *Geocarto International, 33*(3), 240–259.

Singh, R. P., Kumar, V., & Srivastav, S. K. (1990). Technical note Use of microwave remote sensing in salinity estimation. *International Journal of Remote Sensing, 11*(2), 321–330.

Sommer, M., Wehrhan, M., Zipprich, M., Weller, U., Zu Castell, W., Ehrich, S., et al. (2003). Hierarchical data fusion for mapping soil units at field scale. *Geoderma, 112*(3–4), 179–196.

Song, X., Zhang, G., Liu, F., Li, D., Zhao, Y., & Yang, J. (2016). Modeling spatio-temporal distribution of soil moisture by deep learning-based cellular automata model. *Journal of Arid Land, 8*(5), 734–748.

Sonobe, R., Tani, H., Wang, X., Kobayashi, N., & Shimamura, H. (2014a). Random Forest classification of crop type using multi-temporal TerraSAR-X dual-polarimetric data. *Remote Sensing Letters, 5*(2), 157–164.

Sonobe, R., Tani, H., Wang, X., Kobayashi, N., & Shimamura, H. (2014b). Parameter tuning in the Support Vector Machine and Random Forest and their performances in cross- and same-year crop classification using TerraSAR-X. *International Journal of Remote Sensing, 35*(23), 7898–7909.

Sonobe, R., Tani, H., & Wang, X. (2017). An experimental comparison between KELM and CART for crop classification using Landsat-8 OLI data. *Geocarto International, 32*(2), 128–138.

Srivastava, H. S., Patel, P., Navalgund, R. R., & Sharma, Y. (2008). Retrieval of surface roughness using multi-polarized Envisat-1 ASAR data. *Geocarto International, 23*(1), 67–77.

Srivastava, P. K., Han, D., Ramirez, M. R., & Islam, T. (2013). Machine learning techniques for downscaling SMOS satellite soil moisture using MODIS land surface temperature for hydrological application. *Water Resources Management, 27*(8), 3127–3144.

Stisen, S., Sandholt, I., Nørgaard, A., Fensholt, R., & Jensen, K. H. (2008). Combining the triangle method with thermal inertia to estimate regional evapotranspiration—Applied to MSG-SEVIRI data in the Senegal River basin. *Remote Sensing of Environment, 112*(3), 1242–1255.

Stum, A. K. (2010). Random Forests Applied as a Soil Spatial Predictive Model in Arid Utah. Utah State University.

Sun, L., Sun, R., Li, X., Liang, S., & Zhang, R. (2012). Monitoring surface soil moisture status based on remotely sensed surface temperature and vegetation index information. *Agricultural and Forest Meteorology, 166–167*, 175–187.

Sutton, C. D. (2004). Classification and regression trees, bagging, and boosting. *Handbook of Statistics, 24*(04), 303–329.

Sutton, J. R. P., & Lakshmi, V. (2017). From space to the rocky intertidal: Using NASA MODIS sea surface temperature and NOAA water temperature to predict intertidal logger temperature. *Remote Sensing, 9*(162), 1–14.

Taghizadeh-Mehrjardi, R. (2015). Digital mapping of cation exchange capacity using genetic programming and soil depth functions in Baneh region, Iran. *Archives of Agronomy and Soil Science, 62*(1), 109–126.

Taghizadeh-Mehrjardi, R., Sarmadian, F., Minasny, B., Triantafilis, J., & Omid, M. (2014). Digital mapping of soil classes using decision tree and auxiliary data in the Ardakan Region, Iran. *Arid Land Research and Management, 28*(2), 147–168.

Tan, C. P., Ewe, H. T., & Chuah, H. T. (2011). Agricultural crop-type classification of multi-polarization SAR images using a hybrid entropy decomposition and support vector machine technique. *International Journal of Remote Sensing, 32*(22), 7057–7071.

Trenberth, K. E., Fasullo, J. T., & Kiehl, J. (2009). Earth's global energy budget. *Bulletin of the American Meteorological Society, 90*(3), 311–323.

Triantafilis, J., Earl, N., & Gibbs, I. (2012). Digital soil-class mapping across the Edgeroi district using numerical clustering and gamma-ray spectrometry data. In A. B. McBratney (Ed.), *Digital soil assessments and beyond* (pp. 187–191). Sydney: CRC Press.

Tumer, K., & Oza, N. C. (2003). Input decimated ensembles. *Pattern Analysis and Applications, 6*(1), 65–77.

Ulaby, F. T., Bradley, G. A., & Obson, M. C. (1979). Microwave backscatter dependence on surface roughness, soil moisture, and soil texture: Part II-vegetation covered soil. *IEEE Transactions on Geoscience Electronics, 17*(2), 33–40.

Ulaby, F. T., Razani, M., & Dobson, M. C. (1983). Effects of vegetation cover on the microwave radiometric sensitivity to soil moisture. *IEEE Transactions on Geoscience and Remote Sensing, GE-21*(1), 51–61.

Vapnik, V. (1998). *Statistical learning theory*. New York: Wiley-Interscience.

Vapnik, V. N. (2000). The nature of statistical learning theory. In M. Jordan, S. L. Lauritzen, J. F. Lawless, & V. Nair (Eds.), *Statistics for engineering and information science* (pp. 1564–1564). New York: Springer.

Vaysse, K., & Lagacherie, P. (2015). Evaluating digital soil mapping approaches for mapping GlobalSoilMap soil properties from legacy data in Languedoc-Roussillon (France). *Geoderma Regional, 4*, 20–30.

Verhoest, N. E. C., Lievens, H., Wagner, W., Álvarez-Mozos, J., Moran, M. S., & Mattia, F. (2008). On the soil roughness parameterization problem in soil moisture retrieval of bare surfaces from synthetic aperture radar. *Sensors, 8*(7), 4213–4248.

Wagner, W., Lemoine, G., & Rott, H. (1999). A method for estimating soil moisture from ERS scatterometer and soil data. *Remote Sensing of Environment, 70*(2), 191–207.

Wang, L., & Qu, J. J. (2009). Satellite remote sensing applications for surface soil moisture monitoring: A review. *Frontiers of Earth Science in China, 3*(2), 237–247.

Wang, L., Young, S. S., Wang, W., Ren, G., Xiao, W., Long, Y., et al. (2016) Conservation priorities of forest ecosystems with evaluations of connectivity and future threats: Implications in the Eastern Himalaya of China. *Biological Conservation, 195*, 128–135.

Watham, T., Patel, N. R., Kushwaha, S. P. S., Dadhwal, V. K., & Kumar, A. S. (2017). Evaluation of remote-sensing-based models of gross primary productivity over Indian sal forest using flux tower and MODIS satellite data. *International Journal of Remote Sensing, 38*(18), 5069–5090.

Weigend, A. S., Mangeas, M., & Srivastava, A. N. (1995). Nonlinear gated experts for time series: discovering regimes and avoiding overfitting. *International Journal of Neural Systems, 6*(4), 373–399.

Weihermüller, L., Huisman, J. A., Lambot, S., Herbst, M., & Vereecken, H. (2007). Mapping the spatial variation of soil water content at the field scale with different ground penetrating radar techniques. *Journal of Hydrology, 340*(3–4), 205–216.

Wigneron, J. P., Calvet, J. C., Pellarin, T., Van De Griend, A. A., Berger, M., & Ferrazzoli, P. (2003). Retrieving near-surface soil moisture from microwave radiometric observations: Current status and future plans. *Remote Sensing of Environment, 85*(4), 489–506.

Wirth, L., Rosenberger, A., Prakash, A., Gens, R., Margraf, J. F., & Hamazak, T. (2012). A remote-sensing, GIS-based approach to identify, characterize, and model spawning habitat for fall-run chum salmon in a sub-arctic, glacially fed river. *Transactions of the American Fisheries Society, 141*(5), 1349–1363.

Xin, Q., Broich, M., Zhu, P., & Gong, P. (2015). Modeling grassland spring onset across the Western United States using climate variables and MODIS-derived phenology metrics. *Remote Sensing of Environment, 161*, 63–77.

Yang, S., & Huang, Y. (2002). Application of support vector machine based on time series for soil moisture and Nitrate Nitrogen Content prediction. In D. Li & Z. Chunjiang (Eds.), *IFIP advances in information and communication technology* (pp. 2037–2045). Boston: Springer.

Young, S. S. (2003). Satellite detected broad-scale vegetation change in China, 1982–1999. *Asian Geographer, 22*(1–2), 123–142.

Young, S. S., & Harris, R. (2005). Changing patterns of global-scale vegetation photosynthesis, 1982–1999. *International Journal of Remote Sensing, 26*(20), 4537–4563.

Zaman, B., McKee, M., & Neale, C. M. U. (2012). Fusion of remotely sensed data for soil moisture estimation using relevance vector and support vector machines. *International Journal of Remote Sensing, 33*(20), 6516–6552.

Zhang, R., Tian, J., Su, H., Sun, X., Chen, S., & Xia, J. (2008). Two improvements of an operational two-layer model for terrestrial surface heat flux retrieval. *Sensors, 8*(10), 6165–6187.

Zhang, D., Zhang, W., Huang, W., Hong, Z., & Meng, L. (2017). Upscaling of surface soil moisture using a deep learning model with VIIRS RDR. *ISPRS International Journal of Geo-Information, 6*(5), 130.

Zhao, T. J., Zhang, L. X., Shi, J. C., & Jiang, L. M. (2011). A physically based statistical methodology for surface soil moisture retrieval in the Tibet Plateau using microwave vegetation indices. *Journal of Geophysical Research, 116*(D8), D08116.

Zhu, A. X. (2000). Mapping soil landscape as spatial continua: The neural network approach. *Water Resources Research, 36*(3), 663–677.

Zhu, A. X., Hudson, B., Burt, J., Lubich, K., & Simonson, D. (2001). Soil mapping using GIS, expert knowledge, and fuzzy logic. *Soil Science Society of America Journal, 65*(5), 1463.

Chapter 3
Selection of Suitable Variables and Their Development

3.1 Introduction

Model performance depends on the variables used to represent soil–landscape relationships. Therefore variable selection is an important aspect of digital mapping (McKenzie and Ryan 1999; Yang and Huang 2002; McBratney et al. 2003; Finke 2012; Vasques et al. 2012; Vasenev et al. 2014; Xiong et al. 2014; Heung et al. 2016). Variable selection is based on literature survey on crop mapping, digital soil mapping, and soil moisture study using optical or/and microwave satellite data. These variables are put in six classes, i.e. soil properties and parent material; climate; organisms/vegetation; relief/topography; relative position/spatial context; soil moisture and surface characteristics. The variable's categorization is based on 'scorpan' model. McBratney et al. (2003) proposed 'scorpan' model to relate soil class with soil-forming factors considering seven covariates categories ('c' climate; 'o' organisms; 'r' topography; 'p' parent material; 'a' time or age; 's' soil or soil properties, and 'n' relative position). The 'a' age or time of entire surfaces is assumed the same in this study. It relates soil class to soil-forming factors. Soil-landscape parameters are closely intercorrelated, therefore, this study has selected these covariates categories to represent soil–landscape relationships.

Thirty five digital layers are prepared to represent soil–landscape relationships in the prospect of crop cover, soil class, and soil moisture (Table 3.1). In addition to climatic data (precipitation and temperature), a different satellite data from optical to microwave, satellite product, are used to prepare these digital layers.

P. K. Garg et al., *Digital Mapping of Soil Landscape Parameters*, Studies in Big Data 72,
https://doi.org/10.1007/978-981-15-3238-2_3

Table 3.1 Predictor variables associated with six covariates categories

Covariates category	Source data	Intermediate data	Variable layers
Soil properties & parent material	Landsat 8		Band 2 Reflectance
			Band 3 Reflectance
			Band 4 Reflectance
			Band 5 Reflectance
			Band 6 Reflectance
			Band 7 Reflectance
	Landsat 8		Normalized Difference Ratios:
			Bands (3−7)/(3 + 7)
			Bands (4–2)/(4 + 2)
			Bands (5–6)/(5 + 6)
			Bands (5–7)/(5 + 7)
			Bands (6–2)/(6 + 2)
			Bands (6–3)/(6 + 3)
	DEM		Compound Topographic Index (CTI)
	DEM		Integral Moisture Index (IMI)
Spatial context in soil attributes	DEM	CTI	Filtered CTI (5 × 5)
Soil moisture and surface characteristics	RISAT-1	C-Band	σ^{o}_{RH}, σ^{o}_{RV}
	Sentinel-1A		σ^{o}_{VH}, σ^{o}_{VV}
Relief/Topography	DEM		Elevation
			Slope
			Aspect (-π to π)
Spatial context in topography	DEM	Filtered DEM(11 × 11)	Slope
			Curvature
Climate	Monthly precipitation & temperature		Soil Moisture Regimes (SMR)
			Temperature Annual Range (TAR)
			Total Annual Precipitation (TAP)
			Precipitation Seasonality (PS)
			Mean Annual Temperature (MAT)
			Isothermality

(continued)

Table 3.1 (continued)

Covariates category	Source data	Intermediate data	Variable layers
Organism	Landsat 8		NDVI: Bands (5–4)/(5 + 4)
	MODIS NDVI at 16 days Interval (MOD13Q1)		NDVI Variance (Phenological Variability)
Spatial context in an organism	Landsat 8 NDVI		Variance (3 × 3)
			Diversity (3 × 3)
			Diversity (5 × 5)

3.2 Satellite Data and Pre-Processing

3.2.1 Landsat-8

Landsat data are incorporated to generate vegetation, soil, and parent materials variables. One cloud-free Landsat-8 image is retrieved from USGS (United States Geological Survey) Earth Explorer (USGS c2011–2016). The entire study area is covered by a single image. Data acquisition date is 9 March, suitable for *rabi* crops (October–March). The data type of acquired image is USGS L1T, which is georectified and terrain-corrected. Medium resolution of Landsat image (30 × 30 m) is acceptable for crop mapping (Beltran and Belmonte 2001; Ippoliti-Ramilo et al. 2003; Seto and Kaufmann 2005; Alrababah and Alhamad 2006; Bocco et al. 2014; Fan et al. 2014).The Landsat-8 bands 2–7 are initially converted to TOA (Top of Atmosphere) reflectance using a method described in USGS Landsat-8 Product Guide (USGS Landsat-8 2016) and then converted to surface reflectance using Dark Object Subtraction 1 (DOS1) method (Chávez 1996).

3.2.2 Modis

MODISNDVI product MOD13Q1 of the entire year (January–December 2015, 21 images) are retrieved from the United States Geological Survey (USGS) Earth Explorer. Global MOD13Q1 data are provided every 16 days at 250-m spatial resolution. This product is generated using the daily MODIS level-2G surface reflectance. It is a pre-processed data. The MODIS VI algorithm applies a filter to the data based on quality, cloud, and viewing geometry. Cloud-contaminated pixels and extreme off-nadir sensor views are considered lower quality. A cloud-free, nadir view pixel with no residual atmospheric contamination represents the best quality pixel. Only the higher quality, cloud-free, filtered data are used to generate MOD13Q1.

3.2.3 *Digital Elevation Model (DEM)*

Topographic, soil, and parent materials variables are generated by DEM. The 30 m DEM is obtained from ALOS (Advanced Land Observing Satellite) Global Digital Surface Model 'ALOS World 3D—30 m' (© JAXA). The horizontal resolution is ~30 m (1 arcsec) and height accuracy is 5 m as standard deviation (1 sigma). The data set is provided by the Japan Aerospace Exploration Agency (JAXA). The ALOS World 3D—30 m data set is published based on the DSM (Digital Surface Model) data set (horizontal resolution 5 m) of the 'World 3D Topographic Data', which is the most precise global-scale elevation data at this time (Tadono et al. 2014; Takaku et al. 2014).

3.2.4 *Synthetic Aperture Radar (SAR)*

SAR data are also acquired of date 9 March 2015. C-band SAR data of Indian satellite RISAT (Radar Imaging Satellite)-1 in RH (Right circular transmit and Horizontal receive), RV (Right circular transmit and Vertical receive) and Sentinel-1A of ESA (European Space Agency) in VH (Vertical transmit and Horizontal receive), VV (Vertical transmit and Vertical receive) are included in this study. The level 2 CEOS format data in MRS (medium resolution SCANSAR) mode of RISAT-1 is provided by the NRSC (National Remote Sensing Centre) which is georeferenced, terrain-corrected, and right circular polarized data. The spatial resolution of the RISAT-1 image is 18 m with 115 km swath. Radiometric calibration of RISAT-1 product is performed (Mishra et al. 2014). SARC-View software package, developed at Advanced Image Processing Division (AIPD), Space Application Centre (SAC), Indian Space Research Organization (ISRO) has been used to convert RISAT-1 intensity images to backscattering coefficient images. Sentinel-1A data are retrieved from Distributed Active Archive Centres (DAAC) operated by the Alaska Satellite Facility (ASF) (ASFDAAC 2016). Available data for the study area is acquired at Interferometric Wide Beam (IW) mode with 250 km Swath and 5 × 20 m spatial resolution. It is Level-1 Ground Range Detected (GRD) product with dual polarization (VH, VV). Radiometric calibration is performed using SNAP-2.0 (Sentinel Application Platform) software.

3.3 Predictor Variables and Their Development

The predictor variables associated with six categories, i.e. soil properties and parent material, climate, organisms/vegetation, relief/topography, relative position/spatial context, soil moisture, and surface characteristics, are developed using climatic data (precipitation and temperature), a different satellite data from optical (ALOS DEM, Landsat-8) to microwave (RISAT-1, and Sentinel-1A), and satellite product (MODIS NDVI).

3.3.1 Soil Properties and Parent Material

The soil is a combination of inorganic solids, organic matter, gases, and soil water constituents. The soil in any region is the result of the combined activity of environment and terrain features. Different combination and ratio of constituents influence the way in which electromagnetic radiation interacts with the soil surface. Different parent material produced different soil. Shale and phyllite produced silt and clay while quartzite and sandstone produced sand, in combination, they formed soil of different textures according to the different percent of clay, silt, and sand contrasts in the soil. Parent material is the most important factor for controlling soil texture.

Landsat spectral bands and band ratios are effective in distinguishing different geological units (Hunt 1977; El Rakaiby et al. 1994). Landsat-TM band ratio (green/red, red/NIR, NIR/SWIR1) can be used reliably to distinguish white sandstone from other sandstone varieties as well as other sedimentary rock types such as limestone, dolomite, kaolin, and shale (El Rakaiby et al. 1994). Spectral bands around mid-infrared (8–13 μm) are useful to identify the amount of silica content in soil and to discriminate quartzite, carbonate rocks, quartz latitic and quartz monzonitic rocks, latitic and monzonitic rocks, silicified altered rocks, argillized altered rocks, and vegetation (Kahle and Rowan 1980). Normalized difference ratio of the Shortwave Infrared Band (SWIR1) and green band of Landsat ETM+ are useful to identify limestone outcroppings (Bodily 2005). The SWIR1 band appeared to be correlated with the occurrence of near-surface secondary gypsum (Nield et al. 2007). Simple band ratios (red/NIR, red/SWIR1, and SWIR1/SWIR2) with DEM derived data are used to predict soil types (Nield et al. 2007; Dewitte et al. 2012). The use of band ratios also mitigates the influence of shadows. Distinct vegetation communities and landforms are also easily identified in different spectral bands as well as in different normalized difference band ratios (El Rakaiby et al. 1994; McBratney et al. 2003; Bodily 2005; Scull et al. 2005; Nield et al. 2007; Dewitte et al. 2012; Mwaniki et al. 2015; Allbed et al. 2017). However often satellite data need to be combined with ancillary data and field observations in order to be effective (Dewitte et al. 2012).

Furthermore, Compound Topographic Index (CTI) is highly correlated with several soil attributes such as horizon depth ($r = 0.55$), silt percentage ($r = 0.61$), organic matter content ($r = 0.57$), phosphorus ($r = 0.53$), and soil moisture (Moore et al. 1993; Gessler et al. 1995). The terrain analysis software, TauDEM developed by Hydrology Research Group of Utah State University, is used to fill sinks in the DEM data sets, considering only four-way neighbours (TauDEM 1991). D8 and D-infinite algorithm available in TauDEM are used to calculate topographic parameters. The D8 algorithm produces very unrealistic results (Fairfield and Leymarie 1991; Quinn et al. 1991; Costa-Cabral and Burges 1994). Therefore in this work, the

D-infinite algorithm is adopted to calculate slope and flow direction (Tarboton 1997). Compound Topographic Index (CTI) is derived from the specific catchment area (contributing area), flow direction, and slope raster (Moore et al. 1991). Additional filtered CTI layer is also produced using a 5 × 5 low pass filter to introduce spatial context. Integral Moisture Index (IMI) is also suitable to describe soil attributes such as soil pH, available nitrogen (Morris and Boerner 1998; Boerner et al. 2000), vegetation composition (Iverson et al. 1997; Hutchinson et al. 1999), and productivity (Iverson et al. 1997). CTI and IMI are derived from the ALOS DEM. In addition to DEM derived CTI and IMI, 5 × 5 low pass filter is used in CTI image to add spatial context. Landsat-8 OLI bands and different band ratios, ALOS DEM derived CTI and IMI, are used to represent variability in soil and parent material (Fig. 3.1).

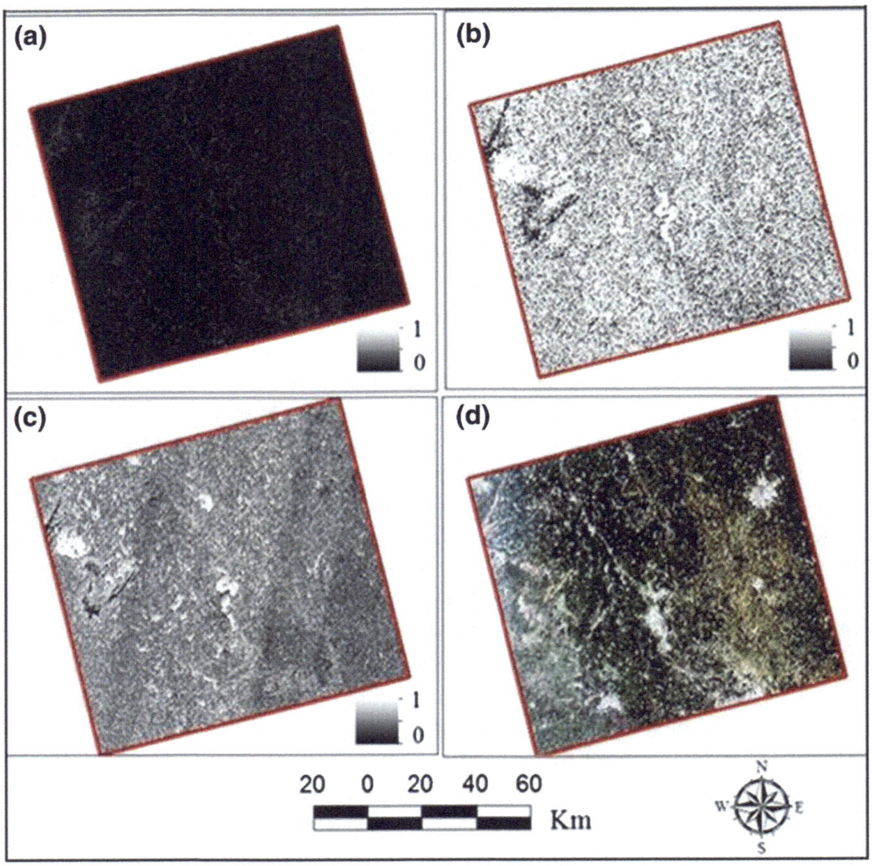

Fig. 3.1 Variables related to soil properties and parent material: **a** IMI; **b** CTI; **c** Filtered CTI; and **d** True colour composite of Landsat-8 bands 4 (red), 3 (green), 2 (blue) images

3.3.2 *Climate*

In the local climate condition, soil-landforms are different according to different climate regimes such as temperature and precipitation (Soil Survey Division Staff 1993). In the high precipitation area, base saturation is less and the process of soil formation is rapid as well as organic carbon present in the soil was removed quickly due to development of soil as well as removal due to erosion whereas in the less precipitation area, base saturation is more. Even in Aridic climate, weathering is governed by diurnal temperature variation. So, the resultant soil is different. Therefore average monthly temperature and average monthly precipitation of time period 1999–2015 are acquired at 21 locations from Climate Change Knowledge Portal (CCKP) of 'The World Bank Group (WBG)' and from the Indian open data portal 'The Open Government Data Platform (OGDP) of India' (GOI/OGDPI 2015, WBG/CCKP 2015).

3.3.2.1 Soil Moisture Regimes (SMR)

Java Newhall Simulation Model (jNSM) is used for Soil Moisture and Temperature Regimes (SMR and STR) determination (Newhall and Berdanier 1996). Average monthly temperature and average monthly precipitation of time period 1999–2015 are used as input to the model. jNSM is a product of Pennsylvania State University and Natural Resources Conservation Service of United States Department of Agriculture. jNSM predicted single STR class-hyperthermic and two SMR classes—aridic (subclass: weak aridic) and ustic (sub-class: typic tempustic) for the study area. jSMR model predicted the ustic (typic tempustic) class at locations for which soil moisture calendar days are dry in some or all the parts ≥90 cumulative days but not dry in all the parts, > half cumulative days when soil temperature >5 °C, and aridic (weak aridic) class for moist in some or all the parts, >45 but <90 consecutive days when soil temperature >8 °C.

Figure 3.2 shows the soil moisture calendar at two locations, Aligarh (a, ustic: typic tempustic) and Bharatpur (b, aridic: weak aridic). Multiple ring buffer analysis is used to trace the delineation line of SMR classes. Multiple buffers at distances of

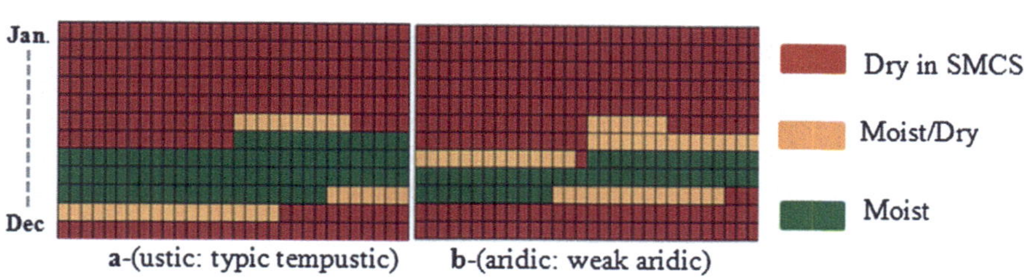

Fig. 3.2 Soil moisture calendar of two classes of SMR at Aligarh [(**a**) ustic: typic tempustic] and Bharatpur [(**b**) aridic: weak aridic]

5000, 6000, 7000, and 8000 m are created around the location points. Intersection points of two classes are joined to trace a boundary of SMR classes. Two polygons are created corresponding to two SMR classes. Raster to vector operation is performed to create SMR raster (Fig. 3.3a).

3.3.2.2 BIOCLIM Parameters

Apart from jNSM, addition to SMR regimes, BIOCLIM model is also used to generate five additional climate parameters, i.e. Mean Annual Temperature (MAT), isothermality, Temperature Annual Range (TAR), Total Annual Precipitation (TAP), and Precipitation Seasonality (PS) (Hijmans et al. 2016). Figure 3.3 shows all climate parameters. The classic climate-envelope-model BIOCLIM of R-package 'dismo' is used to calculate climatic parameters (Hijmans et al. 2016). Monthly climate data namely average monthly maximum, minimum, mean temperature (°C), and precipitation (mm) for 12 months are used as model inputs. Nineteen climate parameters are generated by BIOCLIM model. Only 5 of 19 parameters, namely MAT, isothermality, TAR, TAP, and PS are used in this work (Fig. 3.3b–f). To create raster climate surfaces, Kriging interpolation is applied (Emery 2005).

Figure 3.3a indicated that most of the study area is ustic except the south-western lower part (Bharatpur region), which is aridic. MAT value is also high in the aridic region (Fig. 3.3e); however, it has a moderate value of the 16-year mean precipitation (Fig. 3.3c). This climate zone is suitable for mustard. The standard deviation of the monthly precipitation also ranges from low to moderate in this region (Fig. 3.3d).

3.3.3 Organisms/Vegetation

The vegetation (either natural vegetation or agricultural crops) growing in the region produces a large amount of organic matter and accelerates biological activity. Falling leaves decompose to form the humus and it also accelerates the weathering of primary minerals of soil to form more clay. It also protected the soil from direct rain. It reduces runoff on steep slope thereby allowing more infiltration. Bare soil warms up more quickly and cools off more rapidly. Particularly in hot dry season, the bare soil has diminished biological activity. The production of organic matter is suspended, mineralization of organic matter is intense, and degradation of soil structure to prisms or crusts results. Such soils with degraded top soil layer are more erodible because of low porosity and reduced moisture holding capacity. The remote sensing, Normalized Difference Vegetation Index (NDVI) gives a good representation of vegetation. NDVI is also highly correlated with soil organic carbon distribution. Two first-order texture measures (variance, diversity) of Landsat-8 derived NDVI is also calculated. Two different sizes of neighbourhood windows (3×3, 5×5) are designed to capture the range in the textural measure in NDVI across the study area (Fig. 3.4).

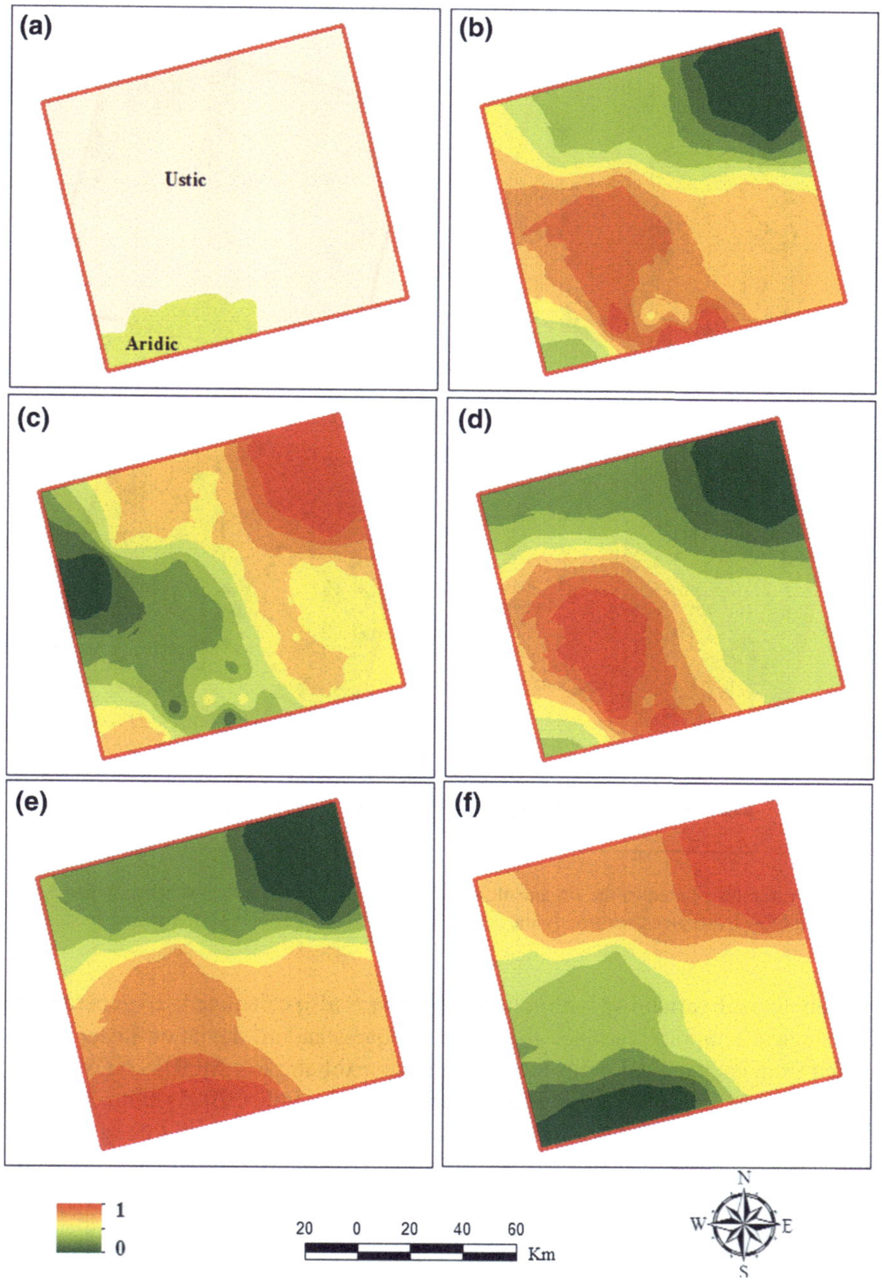

Fig. 3.3 Variables related to climate: **a** SMR; **b** TAR; **c** TAP; **d** PS; **e** MAT; and **f** Isothermality

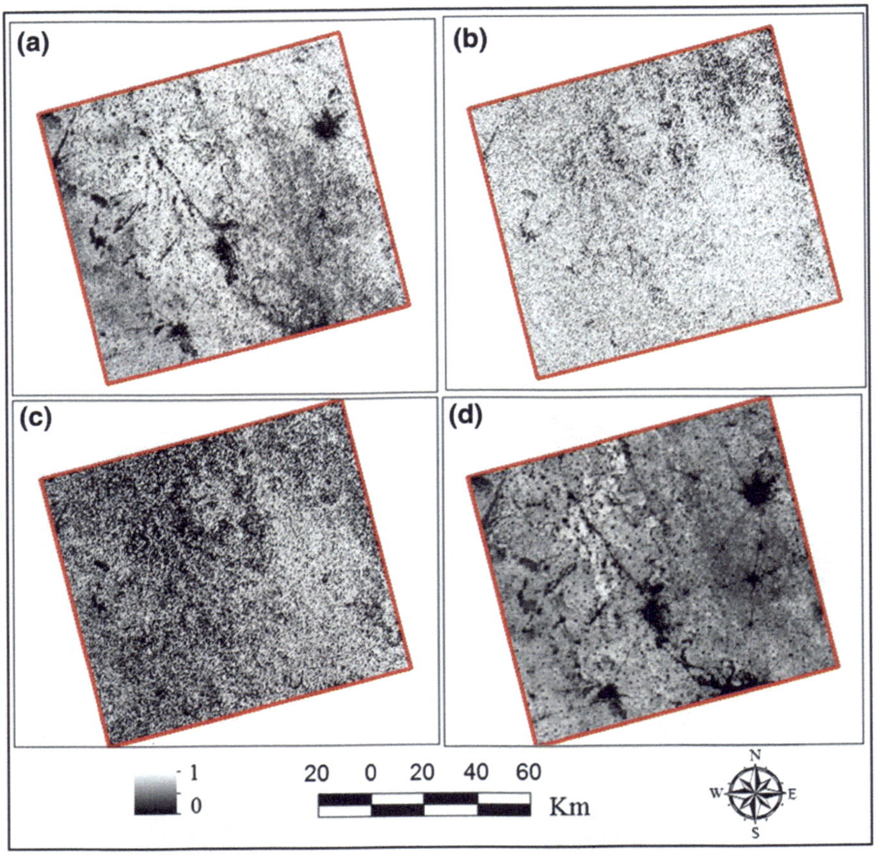

Fig. 3.4 Variables related to the organism: **a** NDVI; **b** NDVI divergence measure by the 3 × 3 window; **c** NDVI variance measure by the 3 × 3 window; and **d** Phenological Variability

Phenological variability within the study area is also calculated. To generate the phenological temporal variance in NDVI, the one-standard-deviation raster is calculated from each annual multi-temporal image pixel-stack of MODIS NDVI product MOD13Q1 of the entire year (January 2015–December 2015) (Fig. 3.4d). In Fig. 3.4d, low standard deviation values correspond to water body (along Yamuna river in study area), urban area (Aligarh, Bharatpur, Hathras, Mathura city, etc.), and fallow land; however, moderate values for forest land and high values for agriculture field (bright colour in image) indicated good suitability of variable for classification.

3.3.4 Relief/Topography

In the northern hemisphere, the southern slopes receive more direct solar radiation than level areas or northern slopes. Hence less evaporation, more soil moisture, more weathering, better vegetation, higher biological activity, and less erosion occurs on northern slopes. Where in the southern slope generally soils are sallow, less developed, more eroded, and have less moisture retention capacity. North- and south-facing slopes exhibit more pronounced differences in soil formation and vegetation cover. Therefore elevation, slope, and aspect are derived from ALOS DEM to represent the topography of the area. A 11 × 11 low pass filter is applied to the 30-m DEM to add spatial context to each pixel (Bodily 2005; Mellor et al. 2013). Slope and curvature for filtered DEM are also calculated (Fig. 3.5). The elevation values range from 90 to 160 m in the study area.

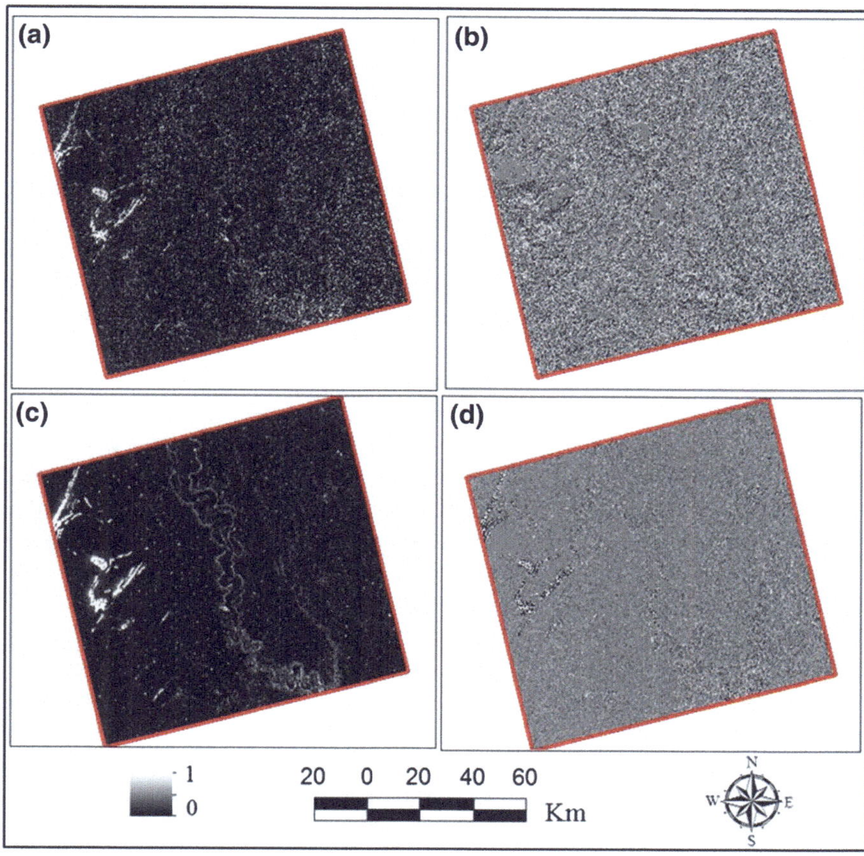

Fig. 3.5 Variables related to relief or topography: DEM derived **a** Slope; **b** Aspect and Filtered DEM derived; **c** Slope; **d** Curvature image

3.3.5 Relative Position

Spatial data analysis seeks to elucidate some pattern or process that occurs in space, perhaps allowing us to make predictions where no observations have been made. Considering a pixel value in context with its neighbours, or its spatial context, can enhance classification (Moran and Bui 2002). To add spatial relationships to the data set, various texture transformations can be performed. Neighbourhood windows of different size are used to add spatial context in the environment covariates related to parent materials and soil, topography, and organism/vegetation. 5×5 low pass filter is used in CTI image to add spatial context (Fig. 3.1e). Two first-order texture measures (variance, diversity) of Landsat-8 derived NDVI is calculated to capture the range in the textural measure in NDVI across the study area as described in (o) organism section (Fig. 3.4b, c). Similarly, a 11×11 low pass filter is applied to the 30-m DEM and slope and curvature is also calculated from filtered DEM to add spatial context in topography, as stated in previous sections (Fig. 3.5c, d).

3.3.6 Soil Moisture and Surface Characteristics

It has been well established and well documented that SAR data are most suitable for remote sensing of soil moisture and to incorporate surface properties like texture, roughness, etc. (Srivastava et al. 2003, 2009; Patel et al. 2006, 2009; Pal et al. 2017). The polarizations are three types, horizontal linear or H, vertical linear or V, and Circular (R: right hand circular, L: left hand circular). Different polarizations mode (HH: horizontal transmit and horizontal receive, VV: vertical transmit and vertical receive, VH: vertical transmit and horizontal receive, HV: horizontal transmit and vertical receive, RH: right hand circular transmit and horizontal receive, RV: right hand circular transmit and vertical receive, LH: left hand circular transmit and horizontal receive, LV: left hand circular transmit and vertical receive) can be used in microwave remote sensing from different platforms (ERS-1/2, JERS-1, SIR-C/X-SAR, RADARSAT-1/2, ENVISAT-1, Sentinel-1, RISAT-1). SAR is sensitive to target properties such as canopy density canopy structure, canopy moisture, soil moisture and surface roughness of the underlying soil (Dobson and Ulaby 1981; Ulaby et al. 1983; Henderson and Lewis 1998; Du et al. 2000; Srivastava et al. 2003, 2009; Patel et al. 2006, 2009; Jia et al. 2012; Pal et al. 2017). These properties are not necessarily specific for the individual classes and caused variability of the backscatter within the classes throughout the study area. However, the sensitivity of SAR backscatter depends upon the frequency, polarization, and angle of incidence at which the SAR is operated (Dobson and Ulaby 1981; Ulaby et al. 1983; Henderson and Lewis 1998; Srivastava et al. 2003; Patel et al. 2006; 2009; Jia et al. 2012). Some study demonstrated that, over bare fields, at low to medium incident angle 20°–37°, like polarizations are more sensitive to soil moisture in comparison to cross polarization (Le Toan et al. 1992; Ranson et al. 1995; Holah et al. 2005; Srivastava et al.

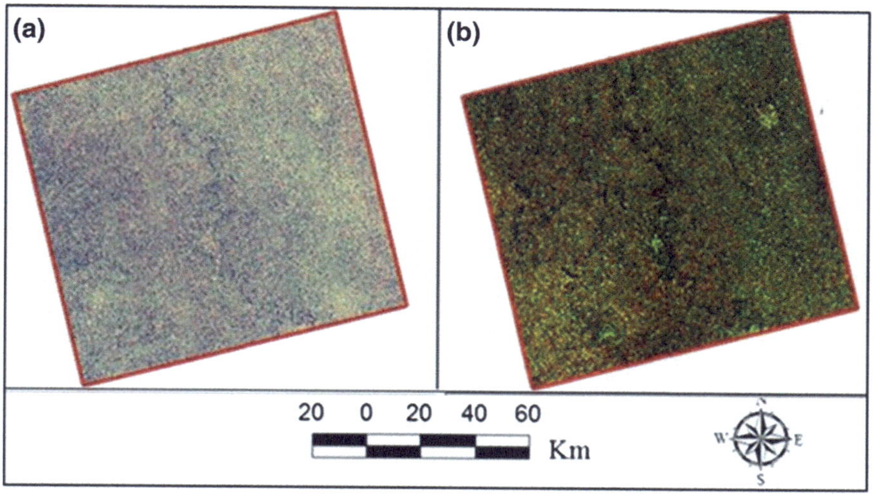

Fig. 3.6 SAR derived data. **a** False colour composite of RISAT-1 polarization bands RH (red), RV (green), 0 (blue); **b** False colour composite of Sentinel-1A polarization bands VH (red), VV (green), 0 (blue)

2006). C-band, HH polarization at relatively steep incidence angles was determined to be the best radar configuration for estimating soil moisture (Ulaby and Batlivala 1976; Ulaby et al. 1979). Similarly, some study observed that the VV polarization are more sensitive to soil moisture compared with the HH polarized (Liu et al. 2016). Other studies found that the sensitivity of the radar signal to soil moisture is lower in HH polarization than in HV polarization for an incidence angle between 20° and 30° (Hirosawa et al. 1978; Ulaby et al. 1979; Kobayashi and Hirosawa 1985; Le Toan et al. 1992). However, results depend on sensor platform, incident angle, target property, polarization types, polarization mode, and wavelength.

Indian SAR satellite RISAT-1 with RH and RV polarization and Sentinel-1A of ESA (European Space Agency) in VH, VV are included in this study (Fig. 3.6). Multi-polarization (cross polarization and like polarization) in two polarization types (linear and circular) are used with other digital data in model development. This study is using C-band SAR at a low incident angle (21°–23°) because, at this frequency and incident angle, the effect of soil roughness and vegetation attenuation is not negligible for surface soil moisture estimation.

All raster layers are resampled to 30 × 30 m pixel size and projected into Universal Transverse Mercator projection, World Geodetic System 1984 datum (UTM 43N, WGS84) to align with each other.

References

Allbed, A., Kumar, L., & Sinha, P. (2017). Soil salinity and vegetation cover change detection from multi-temporal remotely sensed imagery in Al Hassa Oasis in Saudi Arabia. *Geocarto International, 6049*, 1–17.

Alrababah, M. A., & Alhamad, M. N. (2006). Land use/cover classification of arid and semi-arid Mediterranean landscapes using Landsat ETM. *International Journal of Remote Sensing, 27*(13), 2703–2718.

ASF, D. (2016). Alaska satellite facility distributed active archive centers Copernicus Sentinel-1A data center [online]. *Alaska Satellite Facility*. Retrieved November 29, 2016, from https://vertex.daac.asf.alaska.edu/#.

Beltran, C., & Belmonte, A. (2001). Irrigated crop area estimation using Landsat TM imagery in La Mancha Spain. *Photogrammetric Engineering and Remote Sensing, 67*(10), 1177–1184.

Bocco, M., Sayago, S., & Willington, E. (2014). Neural network and crop residue index multiband models for estimating crop residue cover from Landsat TM and ETM+ images. *International Journal of Remote Sensing, 35*(10), 3651–3663.

Bodily, J. (2005). *Developing a digital soil survey update protocol at the golden spike national historic site*. Utah State University Logan.

Boerner, R. E. J., Morris, S. J., Sutherland, E. K., & Hutchinson, T. F. (2000). Spatial variability in soil nitrogen dynamics after prescribed burning in Ohio mixed-oak forests. *Landscape Ecology, 15*(5), 425–439.

Chávez, P. (1996). Image-based atmospheric corrections—revisited and improved. *Photogrammetric Engineering and Remote Sensing, 62*(9), 1025–1036.

Costa-Cabral, M. C., & Burges, J. (1994). Digital Elevation Model Networks (DEMON): A model of flow over hillslopes for computation and dispersal areas specific contributing area. *Water Resources Research, 30*(6), 1681–1692.

Dewitte, O., Jones, A., Elbelrhiti, H., Horion, S., & Montanarella, L. (2012). Satellite remote sensing for soil mapping in Africa: An overview. *Progress in Physical Geography, 36*(4), 514–538.

Dobson, M. C., & Ulaby, F. T. (1981). Microwave backscatter dependence on surface roughness, soil moisture and soil texture: Part III—Soil tension. *IEEE Transactions on Geoscience and Remote Sensing, 19*(1), 51–61.

Du, Y., Ulaby, F. T., & Dobson, M. C. (2000). Sensitivity to soil moisture by active and passive microwave sensors. *IEEE Transactions on Geoscience and Remote Sensing, 38*(1), 105–114.

Emery, X. (2005). Simple and ordinary multigaussian kriging for estimating recoverable reserves. *Mathematical Geology, 37*(3), 295–319.

Fairfield, J., & Leymarie, P. (1991). Drainage network from grid digital elevation models. *Water Resources Research, 27*(5), 709–717.

Fan, C., Zheng, B., Myint, S. W., & Aggarwal, R. (2014). Characterizing changes in cropping patterns using sequential Landsat imagery: An adaptive threshold approach and application to Phoenix Arizona. *International Journal of Remote Sensing, 35*(20), 7263–7278.

Finke, P. A. (2012). On digital soil assessment with models and the Pedometrics agenda. *Geoderma, 171–172*, 3–15.

Gessler, P. E., Moore, I. D., Mckenzie, N. J., & Ryan, P. J. (1995). Soil-landscape modelling and spatial prediction of soil attributes. *International Journal of Geographical Information Systems, 9*(4), 421–432.

GOI/OGDPI. (2015). Government of India/Open Government Data Platform India, transport, socio-economics, climate, agricultural, education and national statics database [online]. *GOI/OGDPI*. Retrieved September 20, 2015, from https://data.gov.in.

Henderson, F.M., & Lewis, J. (1998). Principles and applications of imaging radar. In *Manual of remote sensing* (pp. 461–465). New York: Wiley.

Heung, B., Knudby, A., & Bulmer, C. (2016, December). An overview and comparison of machine-learning techniques for classification purposes in digital soil mapping. *Geoderma, 265*, 62–77.

Hijmans, R.J., Phillips, S., Leathwick, J., & Elith, J. (2016). The comprehensive R archive network package- dismo, species distribution modeling R package.

Hirosawa, H., Komiyama, S., & Matsuzaka, Y. (1978). Cross-polarized radar backscatter from moist soil. *Remote Sensing of Environment, 7*(3), 211–217.

Holah, N., Baghdadi, N., Zribi, M., Bruand, A., & King, C. (2005). Potential of ASAR/ENVISAT for the characterisation of soil surface parameters over bare agricultural fields. *Remote Sensing of Environment, 96*, 78–86.

Hunt, G. R. (1977). Spectral signatures of particulate minerals in the visible and near infrared. *Geophysics, 42*(3), 501–513.

Hutchinson, T. F., Boerner, R. E. J., Iverson, L. R., Sutherland, S., & Sutherland, E. K. (1999). Landscape patterns of understory composition and richness across a moisture and nitrogen mineralization gradient in Ohio (U.S.A.) Quercus forests. *Plant Ecology, 144*(2), 177–189.

Ippoliti-Ramilo, G. A., Epiphanio, J. C. N., & Shimabukuro, Y. E. (2003). Landsat-5 Thematic Mapper data for pre-planting crop area evaluation in tropical countries. *International Journal of Remote Sensing, 24*(7), 1521–1534.

Iverson, L. R., Dale, M. E., Scott, C. T., & Prasad, A. (1997). A GIS-derived integrated moisture index to predict forest composition and productivity of Ohio forests (U.S.A.). *Landscape Ecology, 12*(5), 331–348.

Jia, K., Tian, Y., & Zhang, F. (2012). Crop classification using multi-configuration SAR data in the North China plain. *International Journal of Remote Sensing, 33*(1), 170–183.

Kahle, A. B., & Rowan, L. C. (1980). Evaluation of multispectral middle infrared aircraft images for lithologic mapping in the East Tintic Mountains, Utah. *Geology, 8*(5), 234.

Kobayashi, T., & Hirosawa, H. (1985). Measurement of radar backscatter from rough soil surfaces using linear and circular polarizations†. *International Journal of Remote Sensing, 6*(2), 345–352.

Liu, Y., Zeng, J., Chen, K.-S., & Li, Z. (2016). Parameter sensitivity analysis for bistatic scattering of rough surface. In *2016 IEEE International Geoscience and Remote Sensing Symposium (IGARSS)* (pp. 4335–4338). Beijing, China: IEEE.

McBratney, A. B., Mendonça Santos, M. L., & Minasny, B. (2003). On digital soil mapping. *Geoderma, 117*(1–2), 3–52.

McKenzie, N. J., & Ryan, P. J. (1999). Spatial prediction of soil properties using environmental correlation. *Geoderma, 89*(1–2), 67–94.

Mellor, A., Haywood, A., Stone, C., & Jones, S. (2013). The performance of random forests in an operational setting for large area Sclerophyll forest classification. *Remote Sensing, 5*(6), 2838–2856.

Mishra, M. D., Patel, P., Srivastava, H. S., Patel, P. R., Shukla, A., & Shukla, A. K. (2014). Absolute radiometric calibration of FRS-1 and MRS mode of RISAT-1 Synthetic Aperture Radar (SAR) data using corner reflectors. *International Journal of Advanced Engineering Research and Science, 1*(6), 78–89.

Moore, I. D., Gessler, P. E., Nielsen, G. A., & Petersen, G. A. (1993). Terrain attributes: Estimation methods and scale effects. In A. J. Jakeman, M. B. Beck, & M. J. MCalee (Eds.), *Modelling change in environmental systems* (pp. 189–214). Chichester: Wiley.

Moore, I. D., Grayson, R. B., & Ladson, A. R. (1991). Digital terrain modeling: A review of hydrological geomorphological and biological applications. *Hydrological Processes, 5*(1), 3–30.

Moran, C. J., & Bui, E. N. (2002). Spatial data mining for enhanced soil map modelling. *International Journal of Geographical Information Science, 16*(6), 533–549.

Morris, S. J., & Boerner, R. E. J. (1998). Landscape patterns of nitrogen mineralization and nitrification in southern Ohio hardwood forests. *Landscape Ecology, 13*(4), 215–224.

Mwaniki, M. W., Agutu, N. O., Mbaka, J. G., Ngigi, T. G., & Waithaka, E. H. (2015). Landslide scar/soil erodibility mapping using Landsat TM/ETM+ bands 7 and 3 normalised difference index: A case study of central region of Kenya. *Applied Geography, 64*, 108–120.

Newhall, F., & Berdanier, C. R. (1996). *Calculation of soil moisture regimes from the climatic record.*

Nield, S. J., Boettinger, J. L., & Ramsey, R. D. (2007). Digitally mapping gypsic and natric soil areas using Landsat ETM data. *Soil Science Society of America Journal, 71*(1), 245.

Pal, M., Maity, R., Suman, M., Das, S. K., Patel, P., & Srivastava, H. S. (2017). Satellite-based probabilistic assessment of soil moisture using C-band quad-polarized RISAT1 data. *IEEE Transactions on Geoscience and Remote Sensing, 55*(3), 1351–1362.

Patel, P., Srivastava, H. S., & Navalgund, R. R. (2009). Use of synthetic aperture radar polarimetry to characterize wetland targets of Keoladeo National Park, Bharatpur India. *Current Science, 97*(4), 529–537.

Patel, P., Srivastava, H. S., Panigrahy, S., & Parihar, J. S. (2006). Comparative evaluation of the sensitivity of multi-polarized multi-frequency SAR backscatter to plant density. *International Journal of Remote Sensing, 27*(2), 293–305.

Quinn, P., Beven, K., Chevallier, P., & Planchon, O. (1991). The prediction of hillslope flow paths for distributed hydrological modelling using digital terrain models. *Hydrological Processes, 5*(1), 59–79.

El Rakaiby, M. L., Ashmawy, M. H., Yehia, M. A., & Ayoub, A. S. (1994). In situ reflectance measurements and TM data of some sedimentary rocks with emphasis on white sandstone, southwestern Sinai, Egypt. *International Journal of Remote Sensing, 15*(18), 3785–3797.

Ranson, K. J., Saatchi, S. S., & Sun, G. (1995). Boreal forest ecosystem characterization with SIR-C/XSAR. *IEEE Transactions on Geoscience and Remote Sensing, 33*(4), 867–876.

Scull, P., Franklin, J., & Chadwick, O. A. (2005). The application of classification tree analysis to soil type prediction in a desert landscape. *Ecological Modelling, 181*(1), 1–15.

Seto, K. C., & Kaufmann, R. K. (2005). Using logit models to classify land cover and land-cover change from Landsat Thematic Mapper. *International Journal of Remote Sensing, 26*(3), 563–577.

Srivastava, H. S., Patel, P., Manchanda, M. L., & Adiga, S. (2003). Use of multiincidence angle RADARSAT-1 SAR data to incorporate the effect of surface roughness in soil moisture estimation. *IEEE Transactions on Geoscience and Remote Sensing, 41*(7 PART I), 1638–1640.

Srivastava, H. S., Patel, P., & Navalgund, R. R. (2006). Incorporating soil texture in soil moisture estimation from extended low-1 beam mode RADARSAT-1 SAR data. *International Journal of Remote Sensing, 27*(12), 2587–2598.

Srivastava, H. S., Patel, P., Sharma, Y., & Navalgund, R. R. (2009). Large-area soil moisture estimation using multi-incidence-angle RADARSAT-1 SAR data. *IEEE Transactions on Geoscience and Remote Sensing, 47*(8), 2528–2535.

Tadono, T., Ishida, H., Oda, F., Naito, S., Minakawa, K., & Iwamoto, H. (2014). Precise global DEM generation by ALOS PRISM. In J. Jiang & H. Zhang (Eds.), *ISPRS annals of photogrammetry, remote sensing and spatial information sciences* (pp. 71–76). Suzhou, China: Copernicus Publications.

Takaku, J., Tadono, T., & Tsutsui, K. (2014). Generation of high resolution global DSM from ALOS PRISM. In J. Jiang & H. Zhang (Eds.), *ISPRS—International archives of the photogrammetry, remote sensing and spatial information sciences* (pp. 243–248). Suzhou, China: Copernicus Publications.

Tarboton, D. G. (1997). A new method for the determination of flow directions and upslope areas in grid digital elevation models. *Water Resources Research, 33*(2), 309–319.

TauDEM. (1991). Software package developed by Utah State University-Terrain analysis using digital elevation models.

Le Toan, T., Beaudoin, A., Riom, J., & Guyon, D. (1992). Relating forest biomass to SAR data. *IEEE Transactions on Geoscience and Remote Sensing, 30*(2), 403–411.

Ulaby, F., & Batlivala, P. (1976). Optimum radar parameters for mapping soil moisture. *IEEE Transactions on Geoscience Electronics, 14*(2), 81–93.

Ulaby, F. T., Bradley, G. A., & Obson, M. C. (1979). Microwave backscatter dependence on surface roughness, soil moisture, and soil texture: Part II-vegetation covered soil. *IEEE Transactions on Geoscience Electronics, 17*(2), 33–40.

Ulaby, F. T., Razani, M., & Dobson, M. C. (1983). Effects of vegetation cover on the microwave radiometric sensitivity to soil moisture. *IEEE Transactions on Geoscience and Remote Sensing, GE-21*(1), 51–61.

USGS Landsat-8. (2016). Product guide-using the USGS Landsat 8 Product [online]. *USGS*. Retrieved January 07, 2016, from https://landsat.usgs.gov/using-usgs-landsat-8-product.

Vasenev, V. I., Stoorvogel, J. J., Vasenev, I. I., & Valentini, R. (2014). How to map soil organic carbon stocks in highly urbanized regions? *Geoderma, 226–227*(1), 103–115.

Vasques, G. M., Grunwald, S., & Myers, D. B. (2012). Associations between soil carbon and ecological landscape variables at escalating spatial scales in Florida, USA. *Landscape Ecology, 27*(3), 355–367.

WBG/CCKP. (2015). Climate change knowledge portal of the world bank group [online]. *WBG/CCKP*. Retrieved March 20, 2015, from https://sdwebx.worldbank.org/climateportal/.

Xiong, X., Grunwald, S., Myers, D. B., Kim, J., Harris, W. G., & Comerford, N. B. (2014). Holistic environmental soil-landscape modeling of soil organic carbon. *Environmental Modelling and Software, 57*, 202–215.

Yang, S., & Huang, Y. (2002). Application of support vector machine based on time series for soil moisture and nitrate nitrogen content prediction. In D. Li & Z. Chunjiang (Eds.), *IFIP advances in information and communication technology* (pp. 2037–2045). Boston: Springer.

Chapter 4
Digital Soil Mapping: Implementation and Assessment

As a science grows, its underlying concepts change, although the words remain the same. The following chapter will be devoted to implementation and assessment of DSM. We discuss the various methods that have been, or could be, used for fitting relationships between soil properties or classes and soil-forming factors. We also discussed the data layers that have been, or could be, used to describe the soil-forming factors and their importance in DSM.

4.1 Introduction

Soil plays a vital role in meeting the increasing demand of food and other basic requirements of an ever increasing population of the world. It is now widely recognized that soil survey and soil map is an essential pre-requisite for the planned use of land resources. Emerging countries like Brazil, Russia, India, and China, which are populous, having underdeveloped lands with governments eager to welcome investment from western corporations, are facing double crises because land is limited which could not cope up with the overwhelming population. Hence it is essential to have soil maps for judicious utilization of land, i.e. for the forest, farming, grazing, engineering, etc. (Lookingbill and Urban 2004; Eichhorn 2016).

Soil survey provides relevant information on the types of the soils and their distribution. DSM has advantages over conventional soil mapping approaches as it may better capture observed spatial variability and reduce the need to aggregate soil types based on a set mapping scale (Zhu et al. 2001). Over the last 10 years, DSM has emerged as a credible alternative to traditional soil mapping. However, DSM should not be seen as an end in itself, but rather as a technique for providing data and information for a new framework for soil assessment. Different researchers in DSM have demonstrated that reasonably accurate soil maps can be produced (Shcherbenko et al.

P. K. Garg et al., *Digital Mapping of Soil Landscape Parameters*, Studies in Big Data 72,
https://doi.org/10.1007/978-981-15-3238-2_4

1992; Korolyuk and Shcherbenko 1994; Abdelthaman and Onsi 1996; Connolly and Holden 2009; Panagos et al. 2011; Pásztor et al. 2012; Taghizadeh-Mehrjardi et al. 2014; Taghizadeh-Mehrjardi 2016). DSM approach may also expedite soil survey (Moran and Bui 2002; McBratney et al. 2003; Scull et al. 2005; Cole and Boettinger 2006). Key components of DSM are the 'method' and the set of 'environmental covariates' used to predict soil classes.

Many such methods have been investigated including expert systems (Zhu et al. 2001; Smith et al. 2012; van Zijl et al. 2012), unsupervised classification (Boruvka et al. 2008; Triantafilis et al. 2012), and machine learning. In recent years, machine learning algorithms have received increasing attention for DSM. Several machine learning algorithms have been applied for digital soil mapping of soil types and attributes, such as logistic regression (Hengl et al. 2007; Marchetti et al. 2011; Jafari et al. 2012, 2013; Kempen et al. 2012), classification trees (Lagacherie et al. 1995; Cialella et al. 1997; Bui et al. 1999; McKenzie and Ryan 1999; Moran and Bui 2002; Bui and Moran 2003; Scull et al. 2005; Kim et al. 2012), random forests (Prasad et al. 2006; Grimm et al. 2008; Stum 2010; Poggio et al. 2013; Pahlavan Rad et al. 2014; Vaysse and Lagacherie 2015), neural networks (Grimm et al. 2008; Moonjun et al. 2010; Jafari et al. 2013), and support vector machines (Kovačević et al. 2010). Random Forests (RF) is said to be as accurate as or better than adaptive boosting, yet computationally faster (Breiman 2001). However, the influence of training data quality (noise) and data set size in RF soil classification has not yet been thoroughly investigated (Prasad et al. 2006; Stum 2010; Poggio et al. 2013; Pahlavan Rad et al. 2014). Furthermore, selection of environmental covariates is one of the factors that regulate the performance of machine learning algorithms in DSM (Bui and Moran 2003; Jafari et al. 2013; Poggio et al. 2013).

Beside of that, DSM approach is still unexplored in the Indian context. However, the Indian government is eager to create digital maps of soil type and moved one step forward with The Bill and Melinda Gates Foundation (BMGF) to generate digital maps of soil type and properties of one of its states, Andhra Pradesh (The Hindu 2017). It can be considered as the first step in DSM in India. DSM and space techniques can accelerate the provision of fine scale information on soil fertility and conditions in India. It can also be used in monitoring and understanding soil nutrient change over time.

Considering the fact, in this chapter, the effective framework for implementation of non-parametric algorithms for DSM is suggested. To fulfil the above objective, this study (i) test the performance of RF classifier in comparison with other machine learning methods like CDT (single tree model) and CEB for soil mapping across an arid to ustic area of Indian districts (ii) access the behaviour of RF classifier by considering multiple criteria related to variations in parameter values, and sensitivity to noise and training data size variations. This study utilized 35 digital layers, as described in the previous chapter (Chap. 3), to represent environmental covariates categories, as suggested by McBratney et al. (2003), to present soil–landscape relationship. Potentiality of effective indicators (variables) is analyzed to improve the understanding of the interactive feedback between each parameter and changes of soil and climatic conditions in DSM context. RF grows multiple decision trees as opposed to a single tree in CDT. Thus comparison with CDT allows to evaluate the improvement which is implied by the use of classification ensembles. RF facilitates optimization of two parameters, i.e. choice to select the number of variables at each split (*mtry*) and a number of the tree (*ntree*) to grow thus the comparison with CEB allows to show the effectiveness of RF for soil class mapping.

4.2 Material and Methods

4.2.1 Background to Implementation

4.2.1.1 Preliminary Definitions of Soil

To most people, this is just 'dirt'. To a farmer, it is a potential. To the soil scientist, or pedologist, the word 'soil' conveys a somewhat different meaning, but no generally accepted definition exists (Jenny 1941, 1980).

Hilgard (1914) defined the soil by considering soil primarily as a means of plant production and stated soil as 'The more or less loose and friable material in which, by means of their roots, plants may or do find a foothold and nourishment, as well as other conditions of growth' (Hilgard 1914).

According to Ramann (1911) 'The soil is the upper weathering layer of the solid earth crust' (Ramann 1911). Joffe (1936), a representative of the Russian school of soil science, objects to Ramann's formulation on the grounds that it does not distinguish between soil and loose rock material. According to Joffe, 'The soil is a natural body, differentiated into horizons of mineral and organic constituents, usually unconsolidated, of variable depth, which differs from the parent material below in morphology, physical properties and constitution, chemical properties and composition, and biological characteristics' (Joffe 1936).

4.2.1.2 Soil Formation—A Theoretical Framework

Jenny (1941) correlated the soil properties with independent variables commonly called 'soil-forming factors'. The goal of the functional analysis was the formulation of quantitative laws that permits mathematical treatment to understand the soil differentiations and to explain the geographical distribution of soil types. Parent material (p), climate (c), and organisms (o) are commonly designated as soil formers or soil-forming factors. Since soils change with time and undergo a process of evolution, the factor time (t) also is frequently given the status of a soil-forming factor. Topography/relief (r), which modifies the water relationships in soils and to a considerable extent influences soil erosion, also is usually treated as a soil former. Therefore, Jenny (1941) presented an elegant function to explain the current state of a soil:

$$S = f(\mathrm{c, o, r, p, t}) \tag{4.1}$$

The variables c, o, r, p, and t are the independent variables that define the soil system (Jenny 1941). This equation states that the magnitude of any one of the properties of the 'S (soil)' type such as pH, clay content, porosity, density, and carbonates is determined by the soil-forming factors listed within parentheses. The letter 'f' stands for 'function of', or 'dependent on'. While this function has proven difficult, if not impossible, to solve, it has set forth a theoretical framework whereby soil formation can be studied. To simplify the function and better understand soil formation, studies have focused on identifying transitions in soil properties along sequences of soils related to one environmental factor; such as climate (climosequences), time (chronosequences), and relief (toposequences) (Jenny 1980).

4.2.1.3 Digital Soil Mapping—A Spatial Framework

Spatial data analysis seeks to elucidate some pattern or process that occurs in space, perhaps allowing us to make predictions where no observations have been made. McBratney et al. (2003) proposed a generic framework, which they called the scorpan-SSPFe (soil spatial prediction function with spatially auto correlated errors) method. SSPFe is based on the seven predictive scorpan factors, a generalization of Jenny's five factors, namely, (1) s: soil, other or previously measured attributes of the soil at a point; (2) c: climate, climatic properties of the environment at a point; (3) o: organisms, including land cover and natural vegetation; (4) r: topography, including terrain attributes and classes; (5) p: parent material, including lithology; (6) a: age, the time factor; (7) n: space, spatial or geographic position.

Two additional scorpan covariates are specifically directed towards spatial predictive models: 's' soil or soil properties, and 'n' space, spatial position, or relative position. There are two general forms of the scorpan model (McBratney et al. 2003):

$$S_c = f(\mathrm{s, c, o, r, p, a, n}) \text{ or } S_a = f(\mathrm{s, c, o, r, p, a, n}) \tag{4.2}$$

where S_c is soil class and S_a is a soil attribute or property. Implicit in this are the spatial coordinates x, y, and approximate or vague time coordinate t. Therefore, this can be expressed as

$$\begin{aligned} S_c = f(&\mathrm{s}[x, y, \sim t], \mathrm{c}[x, y, \sim t], \mathrm{o}[x, y, \sim t], \\ &\mathrm{r}[x, y, \sim t], \mathrm{p}[x, y, \sim t], \mathrm{a}[x, y], [x, y]) \end{aligned} \tag{4.3}$$

and a general soil prediction model would be

$$S(x, y, z, t) = f(Q) \tag{4.4}$$

where Q is predictor variable(s) and S stands for $S(x, y, (z), t)$, i.e. the soil class or attribute at some spatial location x, y, (z) and at some time t. The values of predictor variables (independent variables) must be known at the points where the soil class or attributes are to be predicted.

4.2.1.4 Soil-Forming Factors—Environmental Covariates

Most environmental covariates can be represented by remotely sensed spectral data or derivatives from digital elevation models. Satellite imagery and aerial photography are remotely measured properties of the land surface itself, be it soil, water, geology, human infrastructure, or various combinations of these. EM radiation is reflected, absorbed, or emitted as a function of the physical and chemical properties of that surface.

Different satellite bands, ratios, and secondary derivations are used to derive soil-forming factors. Purpose of the first objective was to prepare digital layers representing six classes, i.e. soil properties and parent material; climate; organisms/vegetation; relief/topography; relative position/spatial context; soil moisture and surface characteristics (Table 4.1). These variables selection is based on literature survey on crop mapping, digital soil mapping, and soil moisture study using different satellite and climate data. Different DSM models are reviewed, effective environmental covariates are selected and developed for this study area. The 'a' age or time of entire surfaces is assumed the same in this study. Initially, this study utilized all 35 digital layers (predictor variables) to represent scorpan environmental covariates in the study area. The soil spatial prediction models are developed using these variables.

The DSMs procedure is based on predictive approaches involving environmental covariates, soil reference data. The soil reference data is prior to soil information in point and map form, and field and laboratory observational methods are coupled with spatial and non-spatial soil inference. In this study, pre-existing district Soil Resource Maps (SRM) on 1:50,000 scale of available districts, provided by National Bureau of Soil Survey and Land Use Planning, India are used to obtain the reference data. The maps are resampled (30×30 m) and reclassified in eleven soil classes. These classes are represented by Pedon1-11 throughout the study. 450 soil class samples are collected from each category using stratified random sampling scheme. The number of the training sites per class is kept equal (400 training sites and 50 testing sites per soil category). The structure of DSM and accuracy assessment systems is shown in Fig. 4.1.

4.2.2 Soil Spatial Prediction Models

RF model is tuned by selecting the optimum value of its parameters to get the optimum results. Two other models CDT and CEB are also used. To evaluate the RF technique in a soil classification context, the results achieved by the RF are compared with

Table 4.1 Class wise local importance of variables in descending order of importance

Pedon1	Pedon2	Pedon3	Pedon4	Pedon5	Pedon6	Pedon7	Pedon8	Pedon9	Pedon10	Pedon11
Var. (3)	NDVI	NDVI	Div. (3)	N_Var	Div. (3)	Div. (5)	Div. (5)	Div. (5)	Var. (3)	Div. (5)
NDVI	Div. (5)	Div. (3)	Var. (3)	NDVI	NDVI	F_CTI(5)	CTI	Var. (3)	F_CTI(5)	F_CTI(5)
Div. (5)	Var. (3)	Div. (5)	NDVI	Div. (5)	Var. (3)	NDVI	NDVI	F_CTI(5)	NDVI	Var. (3)
IMI	F_CTI(5)	Var. (3)	Div. (5)	Var. (3)	Div. (5)	Var. (3)	Var. (3)	NDVI	Div. (5)	NDVI
Div. (3)	Div. (3)	N_Var	IMI	Div. (3)	CTI	N_Var	IMI	CTI	IMI	CTI
CTI	N_Var	IMI	F_CTI(5)	IMI	F_CTI(5)	IMI	F_CTI(5)	IMI	N_Var	IMI
F_CTI(5)	IMI	F_CTI(5)	CTI	F_CTI(5)	MAT	CTI	Div. (3)	Div. (3)	Div. (3)	N_Var
N_Var	σ^0RH	CTI	N_Var	CTI	N_Var	σ^0RH	N_Var	N_Var	CTI	Div. (3)
σ^0RV	CTI	PS	σ^0VV	TAP	IMI	Div. (3)	B2	σ^0RH	σ^0VV	σ^0RH
TAP	B2	TAR	TAR	TAR	TAR	TAR	SMR	TAR	TAR	SMR
B2	TAP	SMR	SMR	SMR	SMR	SMR	TAR	SMR	SMR	TAR
Iso	MAT	TAP	TAP	PS	PS	TAP	MAT	MAT	MAT	B2
MAT	PS	MAT	PS	MAT	σ^0RV	σ^0RV	TAP	TAP	σ^0VH	TAP
σ^0VV	σ^0VV	σ^0VH	Iso	σ^0VH	σ^0VV	PS	σ^0VH	σ^0VH	σ^0RV	MAT
σ^0VH	σ^0VH	σ^0VV	σ^0VH	Iso	Iso	σ^0VV	σ^0VV	σ^0RV	TAP	σ^0VV
PS	Asp.	Iso	σ^0RH	σ^0RH	σ^0RH	σ^0VH	Slp	Slp	Asp.	Asp.
SMR	Iso	σ^0RV	Asp.	σ^0VV	Asp.	Iso	PS	Iso	Slp	σ^0RV
σ^0RH	B(4, 2)	B(5, 6)	σ^0RV	Asp.	σ^0VH	Asp.	Iso	σ^0VV	σ^0RH	Slp
Asp.	B(5, 6)	Asp.	B(5, 6)	F_Slp	B(5,6)	Slp	σ^0RV	F_Slp	B(5, 6)	B(5, 6)
B(5, 7)	B(5, 7)	B(5, 7)	F_Slp	σ^0RV	F_Curv	B(5, 6)	Asp.	Asp.	PS	Iso
B(5, 6)	SMR	F_Curv	B(4, 2)	B(5, 7)	Slp	B(4, 2)	B(5, 6)	B(5, 7)	B(4, 2)	B(4, 2)

(continued)

Table 4.1 (continued)

Pedon1	Pedon2	Pedon3	Pedon4	Pedon5	Pedon6	Pedon7	Pedon8	Pedon9	Pedon10	Pedon11
F_Slp	F_Slp	σ^0RH	F_Curv	F_Curv	B(4, 2)	B(5, 7)	B(5, 7)	B(5, 6)	B(3, 7)	σ^0VH
B(4, 2)	F_Curv	B(4, 2)	B(5, 7)	B(5, 6)	B(5, 7)	F_Slp	F_Slp	PS	Iso	B(5, 7)
B(6, 2)	σ^0RV	F_Slp	MAT	B(4, 2)	F_Slp	MAT	F_Curv	B(3, 7)	B(5, 7)	F_Curv
F_Curv	B6	B7	B(3, 7)	B5	TAP	F_Curv	B(4, 2)	B(4, 2)	F_Slp	B(3, 7)
B7	B7	B(3, 7)	B6	B(3, 7)	B5	B6	B5	F_Curv	F_Curv	F_Slp
B6	B(3, 7)	B6	B7	B6	B6	B(3, 7)	B7	B5	B6	B5
B(3, 7)	B(6, 3)	B(6, 3)	Elv	B7	B(3, 7)	B7	B6	B7	B7	B7
B(6, 3)	Slp	Elv	B5	Slp	B7	B5	B(3, 7)	B6	B5	B6
Elv	Elv	Slp	Slp	B(6, 2)	Elv	B(6, 3)	B(6, 2)	Elv	Elv	B(6, 3)
Slp	B(6, 2)	B(6, 2)	B(6, 2)	B2	B(6, 3)	Elv	B(6, 3)	B(6, 2)	B(6, 3)	B(6, 2)
TAR	B3	B3	B(6, 3)	B(6, 3)	B(6, 2)	B(6, 2)	Elv	B(6, 3)	B(6, 2)	Elv
B5	TAR	B4	B3	Elv	B4	B4	σ^0RH	B3	B4	PS
B4	B5	B5	B4	B3	B3	B3	B4	B4	B2	B3

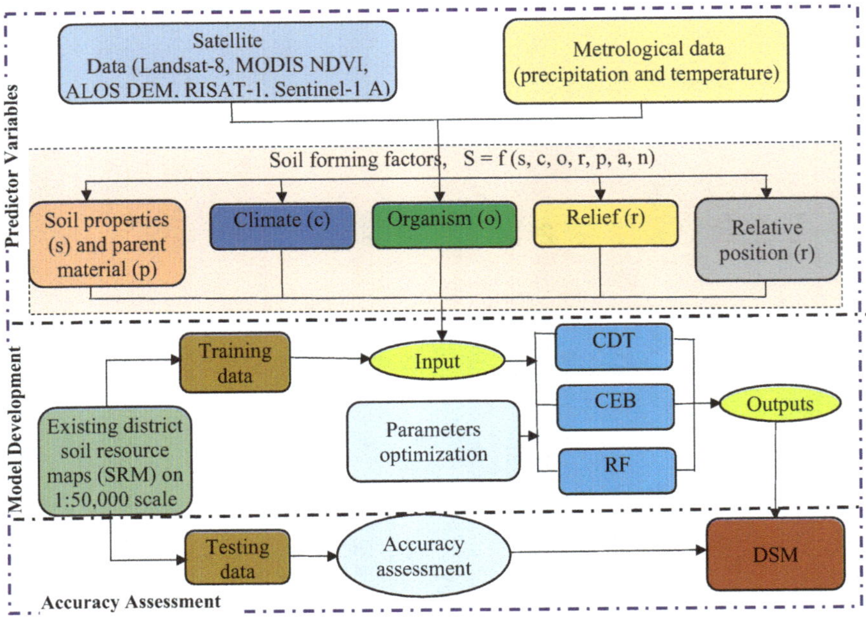

Fig. 4.1 Structure of DSM and accuracy assessment systems

those produced by a CDT and CEB classifiers. The purpose of this comparison is to determine whether the high accuracy values achieved by the RF are algorithm dependent.

Furthermore, machine learning model performance depends strongly on training data quality. Training data containing outliers can reduce model performance. Therefore, outlier analysis is also considered to check the training data quality to ensure that the model performance and behaviour are not influenced by a bad training data set. Generally, if the measure value is greater than 10, it is considered as an outlier.

4.2.2.1 Random Forest Classifier (RF)

RF is a combination of tree predictors such that each tree depends on the values of a random vector sampled independently and with the same distribution for all trees in the forest Breiman (2001). Each tree predictor contributes with a single vote for the assignation of the most frequent class. RF uses the best split of a random subset of predictive variables in the division of every node, which minimize the correlation between the trees in the ensemble. Additionally, in order to increase the diversity of the trees, each tree is constructed using a different bootstrap sample from original data. About one-third of the cases are left out of the bootstrap sample for error estimation, i.e. Out of Bag (OOB). It has proven to be unbiased and accurate in error estimation (Breiman 2001; Rodriguez-galiano et al. 2012). This study has involved

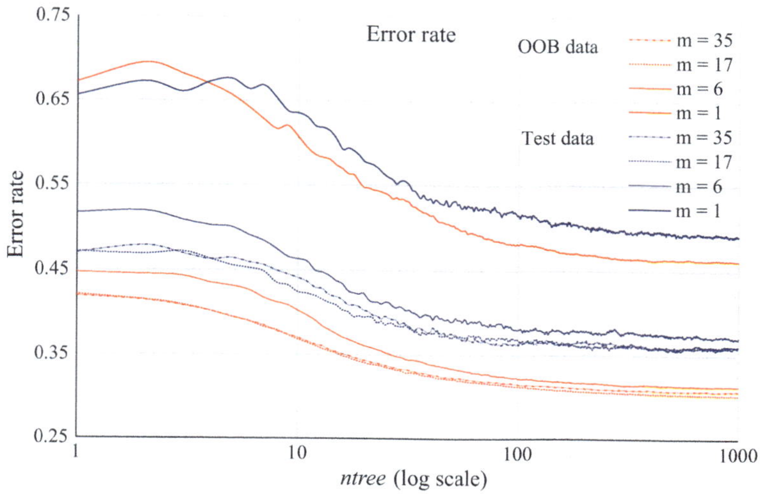

Fig. 4.2 Error rate plot showing dependency on a number of trees and predictive variables

35 variables. It means *mtry* parameter can range from 1 to 35. This value of *mtry* remains constant while the tree is growing, and variable selection is made randomly. Hence the definition of this parameter affects both the correlation and strength of each individual tree and, therefore, it also affects the generalization error and the classifier's accuracy (Breiman 1996, 2001). To understand the effect of parameters *ntree* and *mtry*, error rate is calculated in OOB and test data set for different *mtry*, i.e. D_{min}, D_{max}, D/2, $\sqrt{D}$ with 1000 trees, where D = number of predictor variables (Fig. 4.2).

Selection of the optimal value of *ntree* and *mtry* for RF calibration: The accuracy of classification is the major goal and the speed is the next one. In RF, trees are grown to the maximum potential extent so that they reach the lowest possible bias. RF achieves the lowest error (error = bias + variance) by variance reduction only. However, it is better to have a large number of the tree to reduce the variance but after a certain value, it does not make a difference. A small value of *mtry* will reduce the variance of the ensemble but will also increase the bias of an individual tree and if noisy variables are more, then it will also decrease the probability of choosing an important variable at a split (Louppe et al. 2013; Louppe 2014). Therefore to find the ideal value of parameters *ntree* and *mtry*, Cohen's Kappa Coefficient (CKC) was calculated for *mtry* and *ntree* values at interval 1. A number of trees ranged between 1 to 1000 and number of the variable at the split from 1 to 35. This gave a total number of 35,000 different RF prediction models for the classification and CKC is calculated for each classification. The parameters corresponding to highest CKC value, and lowest *mtry* value are considered as ideal values. Figure 4.3a indicated that for this study, the highest CKC value corresponding to lowest *mtry* is for *mtry* = 6 and *ntree* = 650. Therefore this study has selected *mtry* = 6 and *ntree* = 650 as an optimum value of parameters for DSM of the area using RF classifier. Figure 4.3a

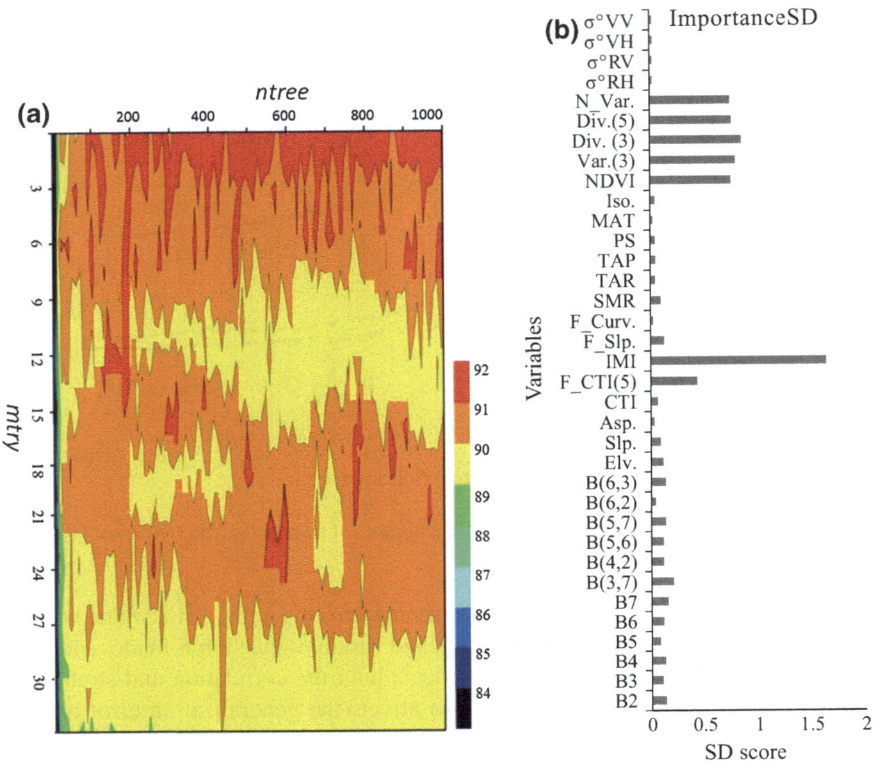

Fig. 4.3 **a** Kappa coefficient plot for *mtry* and *ntree* values; **b** Importance measure on the basis of SD score

shows the relationship between *mtry*, *ntree,* and classification accuracy in terms of CKC.

Variable importance measures: RF gives estimates of what variables are important in the classification. To access the importance of variables to discriminate soil classes, Mean Decrease in Accuracy (MDA) and Mean Decrease Gini (MDG) score have been calculated (Fig. 4.4). The first measure is computed from permuting OOB data: For each tree, the error rate on the OOB portion of the data is recorded. Then the same is carried out after permuting each predictor variable. The difference between the two is then averaged over all trees, and normalized by the standard deviation of the differences. The MDA is based on the hypothesis that if the variable is not important, then permuting the value of that variable will not degrade prediction accuracy (Genuer et al. 2010). The second measure is the total decrease in node impurities from splitting on the variable, averaged over all trees. The MDG ranks variables according to how efficiently the variable separated the classes when selected at a node. In both measurements, permutation score is an average over trees, so there is also importance measurement, SD score, based on standard deviation over trees

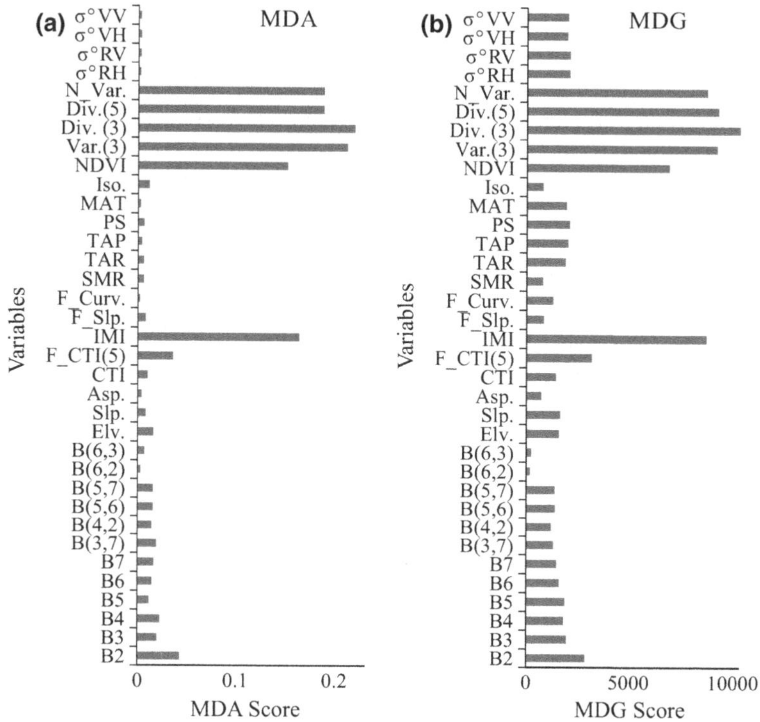

Fig. 4.4 Variable importance contribution of different bands in terms of **a** MDA and **b** MDG

(Fig. 4.3b). Local importance of each class is also measured to know, which, when, and how predictor variable is important to discriminate soil classes (Table 4.1).

4.2.2.2 CART Decision Tree (CDT)

CART algorithm, is a statistical procedure introduced by Breiman et al. (1984), and is used as a classification and regression tool. The CART algorithm provides a foundation for important algorithms like bagged decision trees, RF, and boosted decision trees (Breiman et al. 1984). Single tree model, CDT is used in this study for comparative performance evaluation and behaviour analysis. RF grows multiple trees as opposed to a single tree in CDT model. Therefore, comparison with CDT allows to evaluate the improvement which is implied by the use of classification ensembles.

4.2.2.3 CART Ensemble Bagger (CEB)

Individual decision trees tend to overfit. Bootstrap-aggregated (bagged) decision trees combine the results of many decision trees, which reduces the effects of overfitting and improves generalization. CEB grows the decision trees in the ensemble using bootstrap samples of the data. Bootstrap aggregation, or bagging, is a technique proposed by Breiman (1996) that can be used with many classification methods and regression methods to reduce the variance associated with prediction, and thereby improve the prediction process. RF has the choice to select a number of variables at each split (*mtry*) so that resulting predictions from all of the subtrees (*ntree*) have less correlation (Breiman 1996). However in CEB when selecting a split point, the learning algorithm is allowed to look through all variables in order to select the most optimal split point. Thus even with Bagging, the decision trees can have a lot of structural similarities and in turn, have a high correlation in their predictions. Therefore CEB is also used to compare RF algorithm, which is also an ensemble method with some changes.

4.2.3 Model Construction and Implementation

All modelling was performed in Matlab (MATLAB®). The randomForest-matlab (@misc) package was used to build the RF model and CART algorithm for other methods. The RF code is based on R source by Liaw and Wiener (2002) of RF algorithm, which is originally developed by Leo Breiman and Adele Cutler in FORTRAN (Breiman 2001; Liaw and Wiener 2002). Selection of optimal parameters for RF model calibration is based on the relationship between *mtry*, *ntree,* and the Kappa index. Detailed analysis is presented in Sect. 4.2.2.1. The entire study area was divided into four equal parts to avoid 'out of memory' problem in Matlab. The Dell PC of 6 GB RAM, 500 GB hard disk, and Intel Core i3 processor was used for processing in this study. RF model was implemented independently to predict classes in each part and mosaicked together into a single image. Majority filter (4×4) and expansion–shrinking process were used to remove noise and to smoothen the output. It is assumed that the region of less than four pixels (isolate region) is noise, therefore it is assigned to the nearest class (Fig. 4.5).

4.2.4 Performance Evaluation

4.2.4.1 Accuracy Assessment Measures

Fifty samples of each class are left for the testing purpose. Error matrices are used to assess classification accuracy (Table 4.2). This study adopted popular accuracy measures as suggested by Labatut and Cherifi (2011) and classifiers are evaluated

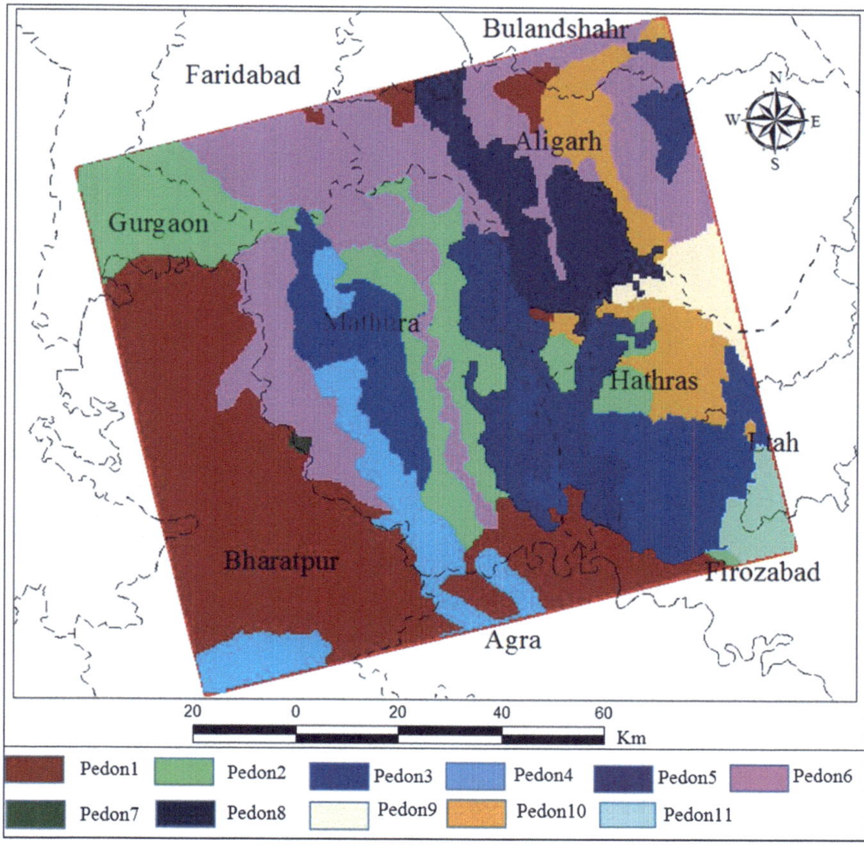

Fig. 4.5 RF predicted soil map of study area

using accuracy measures: (i) overall success rates (OSR), (ii) marginal rates, (iii) F-measure and Jaccard's coefficient of community (Jaccard 1912), (iv) Classification Success Index (CSI), and (v) Agreement Coefficients (AC).

OSR is defined as the trace of the confusion matrix. Under the term marginal rate, reference oriented measures: True Positive Rate (TPR), True Negative Rate (TNR) and prediction oriented measures: Positive Predictive Value (PPV) and Negative Predictive Value (NPV) are calculated (Dice 1945). TPR is also known as recall, sensitivity, producer's accuracy, Dice's asymmetric index and TNR as specificity (Labatut and Cherifi 2011, 2012). PPV is also known as precision, user's accuracy, and Dice's association index (Eqs. 4.5–4.8). All are a function of TP (true positive), TN (true negative), FP (false positive), and FN (false negative).

$$\mathrm{TPR} = \mathrm{TP}/(\mathrm{TP} + \mathrm{FN}) \tag{4.5}$$

$$\mathrm{TNR} = \mathrm{TN}/(\mathrm{TN} + \mathrm{FP}) \tag{4.6}$$

Table 4.2 Performance measures of classifiers RF, CART decision tree, and CART ensembles bagger

		Pedon1	Pedon2	Pedon3	Pedon4	Pedon5	Pedon6	Pedon7	Pedon8	Pedon9	Pedon10	Pedon11	Nor. sco
RF-test	TPR	0.9787^{5}	0.99^{5}	0.9630^{5}	0.96^{5}	0.9688^{5}	0.9320^{5}	0.9333^{5}	0.9394^{5}	0.9667^{5}	0.9275^{5}	0.98^{5}	5.0
	TNR	0.9979^{5}	1^{5}	1^{5}	0.9848^{5}	1^{5}	0.9857^{5}	1^{5}	0.9882^{5}	0.9950^{5}	0.9984^{5}	0.9721^{5}	5.0
	PPV	0.9787^{5}	1^{5}	1^{5}	0.7619^{3}	1^{5}	0.9411^{5}	1^{5}	0.9490^{5}	0.8529^{4}	0.9973^{5}	0.9560^{5}	4.72
	NPV	0.9979^{5}	0.9989^{5}	0.9980^{5}	0.9980^{5}	0.9980^{5}	0.9833^{5}	0.9980^{5}	0.9859^{5}	0.999^{5}	0.9566^{5}	0.9874^{5}	5
	F-meas.	0.9787^{5}	0.9950^{5}	0.9811^{5}	0.8496^{4}	0.9841^{5}	0.9365^{5}	0.9655^{5}	0.9444^{5}	0.9063^{5}	0.9611^{5}	0.9679^{5}	4.91
	JCC	0.9583^{5}	0.99^{5}	0.9630^{5}	0.7385^{3}	0.9688^{5}	0.8807^{4}	0.9333^{5}	0.8942^{4}	0.8286^{4}	0.9252^{5}	0.9378^{5}	4.55
	ICSI	0.9574^{5}	0.99^{5}	0.9630^{5}	0.7219^{3}	0.96875^{5}	0.8732^{4}	0.9333^{5}	0.8883^{4}	0.8198^{4}	0.9248^{5}	0.9360^{5}	4.55
	Nor. scor	5.0	5.0	5.0	4.0	5.0	4.7	5.0	4.7	4.57	5.0	5.0	
	CKC: 0.937^{5}; OSR: 0.9598^{5}; CSI: 0.9070^{5}; Total time: $00:0:59^{5}$												
CDT	TPR	0.9787^{5}	0.945^{5}	0.9630^{5}	0.8571^{4}	0.9688^{5}	0.8878^{4}	0.9333^{5}	0.9192^{5}	0.90^{5}	0.9630^{5}	0.9698^{5}	4.82
	TNR	0.9979^{5}	0.9952^{5}	0.9939^{5}	0.9905^{5}	1^{5}	0.9832^{5}	0.9940^{5}	0.9856^{5}	0.990^{5}	0.9613^{5}	0.9717^{5}	5.0
	PPV	0.9787^{5}	0.9792^{5}	0.8966^{4}	0.9545^{5}	1^{5}	0.9255^{5}	0.8235^{4}	0.9381^{5}	0.9836^{5}	0.5777^{1}	0.9554^{5}	4.45
	NPV	0.9979^{5}	0.9870^{5}	0.9980^{5}	0.9674^{5}	0.9979^{5}	0.9740^{5}	0.998^{5}	0.9810^{5}	0.9402^{5}	0.9978^{5}	0.981^{5}	5.0
	F-meas.	0.9787^{5}	0.9618^{5}	0.9286^{5}	0.9032^{5}	0.9841^{5}	0.9062^{5}	0.875^{4}	0.9286^{5}	0.9399^{5}	0.7222^{3}	0.9626^{5}	4.73
	JCC	0.9583^{5}	0.9264^{5}	0.8667^{4}	0.8235^{4}	0.9688^{5}	0.8286^{4}	0.7778^{3}	0.8667^{4}	0.8867^{4}	0.5652^{1}	0.9279^{5}	4.0
	ICSI	0.9574^{5}	0.9242^{5}	0.8595^{4}	0.8117^{4}	0.9688^{5}	0.8132^{4}	0.7569^{3}	0.8573^{4}	0.8836^{4}	0.5407^{1}	0.9252^{5}	4.0
	Nor. scor	5.0	5.0	4.57	4.57	5.0	4.57	4.14	4.71	4.71	3	5.0	
	CKC: 0.8465^{5}; OSR: 0.9039^{5}; CSI: 0.8453^{4}; Total time: $00:00:40^{5}$												
CEB Bagger	TPR	0.9787^{5}	0.9333^{5}	0.9630^{5}	0.9394^{5}	0.9688^{5}	0.8811^{4}	0.9333^{5}	0.9293^{5}	0.8932^{4}	0.9787^{5}	0.98^{5}	4.57
	TNR	0.9979^{5}	1^{5}	0.9960^{5}	0.9810^{5}	1^{5}	0.9833^{5}	0.9980^{5}	0.9834^{5}	0.9880^{5}	1^{5}	0.9720^{5}	5.0
	PPV	0.9787^{5}	1^{5}	0.9286^{5}	0.9208^{5}	1^{5}	0.9271^{5}	0.9333^{5}	0.9293^{5}	0.9484^{5}	1^{5}	0.9561^{5}	5.0

(continued)

Table 4.2 (continued)

		Pedon1	Pedon2	Pedon3	Pedon4	Pedon5	Pedon6	Pedon7	Pedon8	Pedon9	Pedon10	Pedon11	Nor. sco
	NPV	0.9979^{5}	0.9980^{5}	0.9980^{5}	0.9857^{5}	0.9980^{5}	0.9717^{5}	0.9980^{5}	0.9834^{5}	0.9742^{5}	0.9979^{5}	0.9873^{5}	5.0
	F-meas.	0.9787^{5}	0.9655^{5}	0.9454^{5}	0.93^{5}	0.9841^{5}	0.9036^{5}	0.9333^{5}	0.9293^{5}	0.92^{5}	0.9892^{5}	0.9679^{5}	5.0
	JCC	0.9583^{5}	0.9333^{5}	0.8966^{4}	0.8692^{4}	0.969^{5}	0.8241^{4}	0.875^{4}	0.8679^{4}	0.8519^{4}	0.9787^{5}	0.9378^{5}	4.45
	ICSI	0.9574^{5}	0.9333^{5}	0.8915^{4}	0.8602^{4}	0.9688^{5}	0.8082^{4}	0.8667^{4}	0.8585^{4}	0.8417^{4}	0.9787^{5}	0.9360^{5}	4.45
	Nor. scor	5.0	5.0	4.71	4.71	5.0	4.57	4.71	4.71	4.57	5.0	5.0	
	CKC: 0.9020^{5}; OSR: 0.9281^{5}; CSI: 0.9001^{4}; Total time: $00{:}04{:}40^{1}$												

$$\mathrm{PPV} = \mathrm{TP}/(\mathrm{TP} + \mathrm{FP}) \tag{4.7}$$

$$\mathrm{NPV} = \mathrm{TN}/(\mathrm{TN} + \mathrm{FN}) \tag{4.8}$$

F-measure and JCC can be calculated from Eqs. 4.9 and 4.10:

$$\mathrm{F} = 2(\mathrm{PPV} \times \mathrm{TPR})/(\mathrm{PPV} + \mathrm{TPR}) \tag{4.9}$$

$$\mathrm{JCC} = \mathrm{F}/(2 - \mathrm{F}) \tag{4.10}$$

CSI is an average of individual classification success index, where ICSI = PPV + TPR − 1. Under AC, popular Cohen's Kappa Coefficient (CKC) is calculated for 95% Confidence Interval (CI). TPR, TNR, PPV, NPV, ICSI, JCC, and F-measures are class specific measure while, OSR, CSI, and CKC are multiclass measures. CKC ranges from $-\infty$ to $+1$, ICSI, CSI from -1 to $+1$, and all other measures ranges from 0 to $+1$. A value near 1 indicates good quality of measure. Total time (training + testing time) is also included as one assessment parameter to evaluate the performance of predictive models.

4.2.4.2 Comparative Performance Method

Popular accuracy measures as suggested by Labatut and Cherifi (2011) are used for comparative performance analysis. The grading system is adopted to assign a class value for each measurement (Shukla et al. 2018) (Table 4.3). Each measure value is assigned a class value ranging from 1 to 5. Class value is corresponding to different ranges of accuracy measures. The Score is calculated for each class as well as for measures on the basis of class value. Then score corresponding to each measure is normalized by dividing the number of classes and, score corresponding to each class is normalized by a number of measures. Total time (training + testing time) is also included to evaluate performance. To assign a class value to time, lowest time (TL) is assigned the highest class value (C = 5) and the fraction is calculated for other time value (f = T/TL) to calculate the class value (C/f). The overall performance of methods is evaluated based on total score (Table 4.2).

Table 4.3 Grading of performance

Accuracy measures	Class value	Classifier performance
0.90–1.0	5	Excellent
0.80–0.90	4	Good
0.70–0.80	3	Fair
0.60–0.70	2	Poor
<0.60	1	Less than poor

4.2.5 *Parameters to Access Model Behaviour*

The behaviour of RF is assessed by considering multiple criteria related to variations in classifier parameter values, sensitivity to noise, and training size variations. To evaluate the effect of training size in the model, subsets of original training data obtained by reducing the original training data in increments of 5% are used to train the models. Then CKC value of classification is calculated by testing data set for each model.

To get the noise effect on the model, again CKC value is used for model evaluation. Noise in training set is produced by assigning a false class to the training set in 5% increment and model trained by these noise data to get model accuracy. The behaviour of RF to these parameters is compared with CDT and CEB.

4.3 Results and Discussion

4.3.1 *RF Dependency on Parameters (***mtry, ntree***)*

As explained in Sect. 4.2.2.1, RF brings in an additional parameter which does not appear in traditional classification trees, i.e. *mtry*. Figure 4.2 shows the error depending on the number of trees for different m values. From Fig. 4.2, it can be observed that (i) Average absolute difference between error rate calculated on test and OOB data is very less ($\approx$0.04). Therefore the OOB estimation is as accurate as the estimation based on the test data set (Breiman 1996, 2001). The advantage over test data is that the OOB estimation is unbiased. Therefore in RF, it is not necessary to use an independent test data set, OOB estimation is sufficient to represent model performance. (ii) The average of the absolute differences of the OOB estimation between the minimum and maximum of the possible random variables (1 and 35) from tree number 100 is less than 1%, from which it can be concluded that an RF is not sensitive to the value of *mtry* once the point has been reached in which the error converges. (iii) It can be observed from the plot that for any *mtry* value, change in error rate is measurable upto a certain *ntree* value, after this error rate change is negligible, i.e. classification model is stable after the certain *ntree* value. However, a large number of *trees* increase computational time. Therefore it is required to get the optimum value of *ntree* instead of randomly assigning any large number. (iv) Error rate convergence is low for *mtry* 1 and 6 in comparison to higher *mtry*. However, a large number of *mtry* also increases computational time thus, it is suggested that RF should be optimized to find the optimum *mtry* value.

Section 4.2.2.1 has described the methodology to select the optimum value of parameters to tune the RF. Based on the relationship between *mtry*, *ntree,* and CKC, it is concluded that the relationship between them is directly proportional until a certain number of trees is achieved from which CKC remains stable (Fig. 4.3a). It means an increase in *ntree* or *mtry* after this, only increase computational time without having any contribution on soil classification accuracy. Thus this result also emphasized the previous finding.

4.3.2 Identification of Effective Indicators

Identification of effective indicators are based on variables importance (MDA, MDG, and SD score) (Figs. 4.3b and 4.4). With the high score in both the measures (MDA and MDG), Vegetation related variables: NDVI, texture measure in NDVI (Variance and Diversity), and phenological temporal variance in NDVI (NDVI variance) have shown their importance in DSM approach. These variables have also shown importance in SD score. A large portion of the study area is agricultural land (93%). Many common agricultural practices, especially ploughing, disc-tillage, and vegetation burning, accelerate the decomposition and humification of soil organic matter and leave the soil susceptible to wind and water erosion. Thus they have an influence on soil formation. Therefore, it was expected to get a good score for vegetation related variables in digital soil mapping. Diversity measurement with window size 3 in NDVI is top scorer in MDA and MDG scores. However, it is second most important variable in SD score followed by IMI. Therefore, it indicates that not only the variables associated with vegetation but also their texture measurements are very useful in the soil prediction model.

Topographic variables associated with soil attributes (CTI, IMI) scored well in all scores. IMI is under top five scorers in MDA and MDG score. In SD score, IMI is a top scorer. Predictor variable filtered CTI is at 7th position in all importance measurements. It indicates that potential of CTI and IMI is useful in class prediction in the model. Accumulation numbers in flat areas will be very large and therefore CTI has not shown as much importance as expected. However, DEM derived topographic variables; slope and aspect are not among the top ten variables in any of the important measurements. Therefore, slope and aspect appeared to have little impact on soil class prediction in the relatively flat surface.

Landsat-8 spectral bands in the visible spectral range are in the top ten most important variables in all measurements. However other bands and their ratio are not in the top ten but they have also shown significant importance in all scores. Therefore it also shows a significant impact of spectral response and soil information associated with band ratios in model development.

However, climate variables have not shown significant importance in MDA and SD score but these scored well in MDG measurement. Similarly, the backscattering responses in two polarization modes have also scored lower values in MDA and SD score but comparatively significant in MDG score. Therefore it may be assumed that all predictor variables have an impact on the model and this study has involved all type of variables for model development.

Three types of variables are involved in this study, i.e. spectral, topographic (DEM derived), and climate. Spectral and topographic variables have major importance over climate variables. These variables are used separately to produce a forest of 650 trees with *mtry* value 6 and OOB error rate has been calculated to test the comparative importance of topographic, spectral, and climate variables. OOB error rate for the spectral only is 0.65, for topographic only is 0.69, and for climate only is 0.86. The combination of spectral and topographical variables produced OOB error rate of 0.35 and if climate variables are also involved, then it reduces to 0.32. Thus it is assumed that in the final RF model, it is better to use all these three types of variables instead of eliminating any one.

Local importance of each class is also measured to know, which, when, and how predictor variable is important to discriminate soil classes. Table 4.1 shows the importance of the contribution of each variable to the RF classification of a specific class. In general terms, it can be stated that the class specific behaviour is the same as for the overall contribution as discussed above, although for some categories, the backscattering response and climate variables contribute significantly. It indicates the importance of these variables in class-separability for particular classes. It can also be observed that how the variables derived from the vegetation indices have the strongest influence in the class-separability of the classification scheme.

4.3.3 Accuracy Assessment

Digital soil map of the study area is presented in Fig. 4.5. The results of the accuracy assessment of the map obtained through the RF classification model are very positive, taking into account the complexity of the study area and a large number of categories (Table 4.2). In test data set for Pedon2, 3, 5, PPV and NPV $\approx$1, indicating perfect prediction (i.e. classifier prediction is 100% reliable). All marginal rates scored excellent measures value except Pedon4 (PPV = 0.7619) and Pedon9 (NPV = 0.8529). All other class specific measures are a function of TPR and PPV. For all soil classes, all class specific measures values are under excellent value range except Pedon4 (PPV, F-measure, JCC, ICSI $<$ 0.90), Pedon6, 8, and 9 (JCC, ICSI $<$ 0.90). In the case of Pedon1, 2, 6, and 8 TPR $\approx$ PPV and TNR $\approx$ NPV ($\pm$0.01), it showed the same level of prediction reliability and performance of classifier for these classes. High TPR and TNR values indicate classifier is good at classifying instances belonging to a specific true class, as much for instances belonging to other classes. The high value of other class specific measures, as well as all multiclass measures also (CKC: 0.937; OSR: 0.9598; CSI: 0.907), indicated good performance of the classifier.

4.3.4 Comparative Performance Analysis

Digital soil maps of study area obtained using CDT and CEB are presented in Fig. 4.6. Details of the ranking the method and their accuracy measures are presented in methodology Sect. 4.2.4.

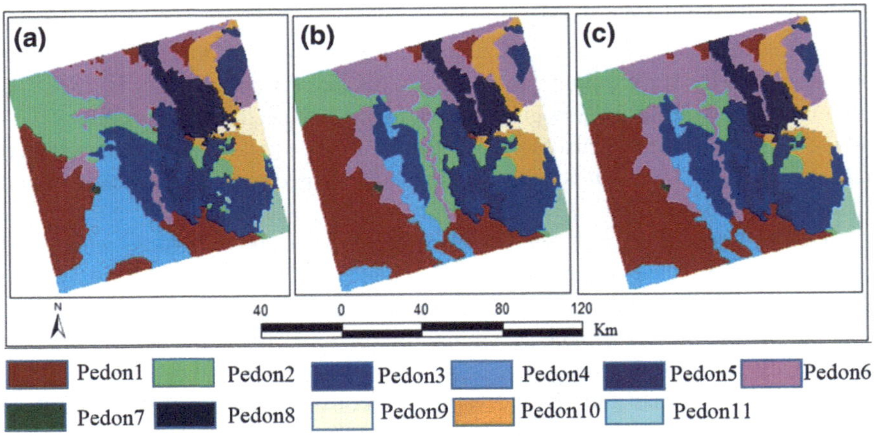

Fig. 4.6 **a** CDT; **b** RF; and **c** CEB predicted soil map of study area

On the class specific measures scale, RF performed well for all soil classes except Pedon4 (Fig. 4.7). From Table 4.2, it can be observed that for all soil classes, marginal rates, as well as other measures value, are high in RF method. Thus, scored well in gradient scaling for all classes except one soil class, Pedon 4. The column corresponding to Pedon 4 shows a low value of JCC and ICSI measures (<75%) in RF model. Both measures depend on True and Predictive Positive Rate (PPV and TPR). In the case of Pedon 4, TPR value is high (96%) but PPV value is low (76%). However, negative rates value is high, it means RF is overfitted for this class, i.e. model performs well on the training data for this class but does not perform well on the evaluation data. However, for all other classes, RF performed well.

All models equally performed for soil classes 1, 2, 5, and 11 with highest score 5.0 (Fig. 4.7). For these soil classes, marginal rates, as well as other measures value, are high in all three models (Table 4.2). Therefore, all methods achieved the highest score for these soil classes in gradient scaling. However, CDT shows the lowest

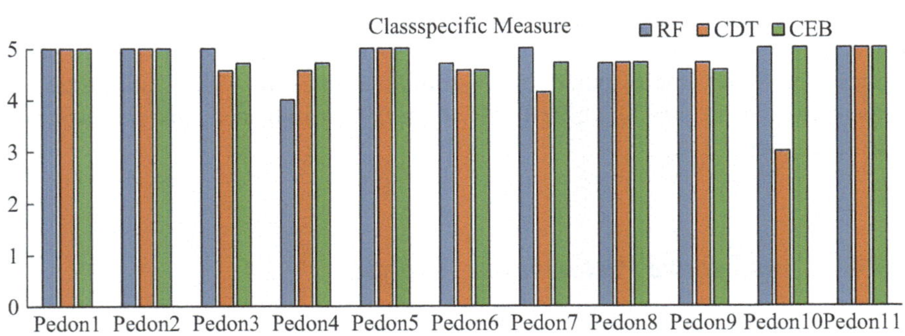

Fig. 4.7 Performance based on class specific measures

performance for soil classes 3, 7 and 10 with low PPV value and indicates overfit of the model for these classes. CEB score is approximately equal to RF for the majority of soil classes. Therefore on the basis of class specific measures scale, it can be stated that ensemble methods are good for DSM for most of the classes.

Furthermore, Fig. 4.8 shows performance of method based on multiclass measures scale. It is the sum of the class value assigned to multiclass measures. It can be observed from the plot that in multiclass measure scale, RF is top scorer. However, CEB and CDT scored same in this scale. Multiclass measures indicate the performance of the model for overall all classes. Therefore performance based on multiclass measures indicates that performance of classifier and prediction reliability is good for soil mapping using RF model.

Time plot is also shown in the same figure with darker boundary in the textured bar (Fig. 4.9). It can be observed that CDT requires less computational time in comparison to RF and CEB. Although, the result is obvious, as CDT is single tree model. For computational load, a single decision tree will train much more quickly, and computing a prediction is also much quicker. But in terms of classification accuracy, ensembles methods are performed better than CDT. In ensemble models, RF model is quicker than CEB, because a constituent tree of RF is different from CEB in the way of splitting criterion.

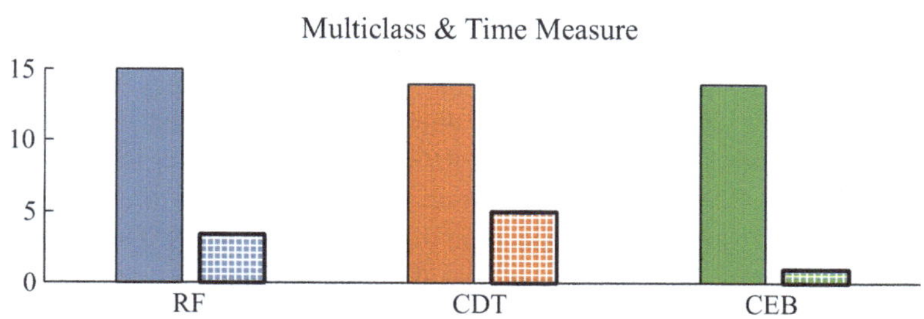

Fig. 4.8 Performance based on multiclass measures and time

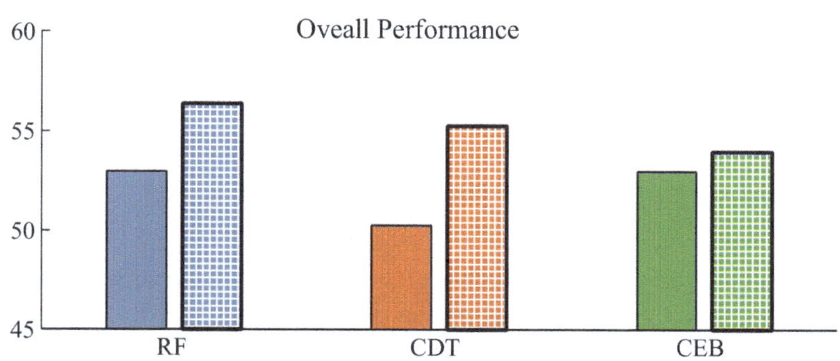

Fig. 4.9 Overall performance based on multiclass measures and time

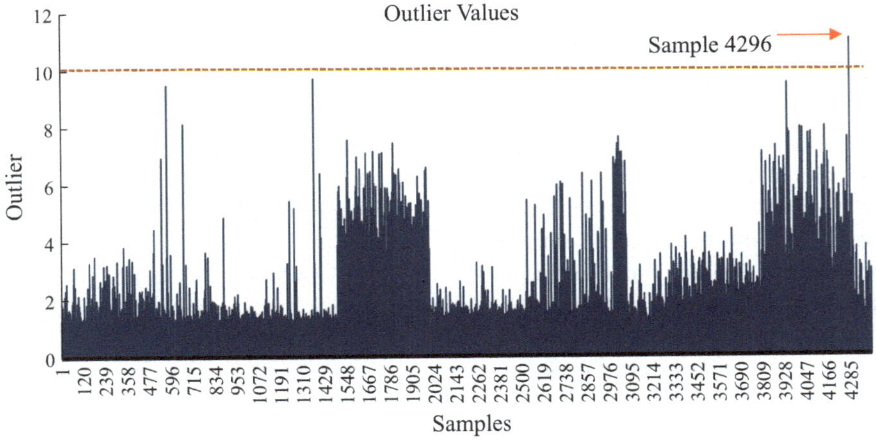

Fig. 4.10 Outlier analysis for each sample

Figure 4.9 shows model ranking based on overall performance of models without and with time factor (bar with dark boundary in the textured bar). Because RF scored highest in both cases, therefore from plot, it can be concluded that RF is a robust method with high classification accuracy. If computational time is not considered as one factor to evaluate the performance of the model, CEB also performed well. It indicates the good mapping capability of bootstrap-ensembling over single tree model. Furthermore, RF is more robust than CEB, as described in the previous discussion. Therefore it can be stated that RF is the best performer for DSM.

From outlier analysis, it is also clear that almost all samples are true samples, therefore accuracy results purely depend on the model (Fig. 4.10). Thus from above discussion, it can also be summarized that the high accuracy values achieved by the RF are algorithm dependent. In outlier, the sample 4296 can be considered as exceptions, which belongs to the Pedon11. However soil class, Pedon11 scored well in class specific measure score. Because machine learning methods have resistance to noise and single noise cannot affect the result (Breiman 1996, 2001; Rodriguez-galiano et al. 2012). Section 4.3.6 is also described more on model resistance to noise.

4.3.5 Influence of Training Data Set Size

The various studies have demonstrated that it is necessary to have larger training data set in order to obtain optimal results in non-parametric machine learning methods (Foody 1995; Foody and Arora 1997; Moran and Bui 2002; Pal 2005; Pal and Mather 2005). Training data should be sufficient to represent all the variability in category across the study area. It is important to plan a sampling scheme to achieve acceptable

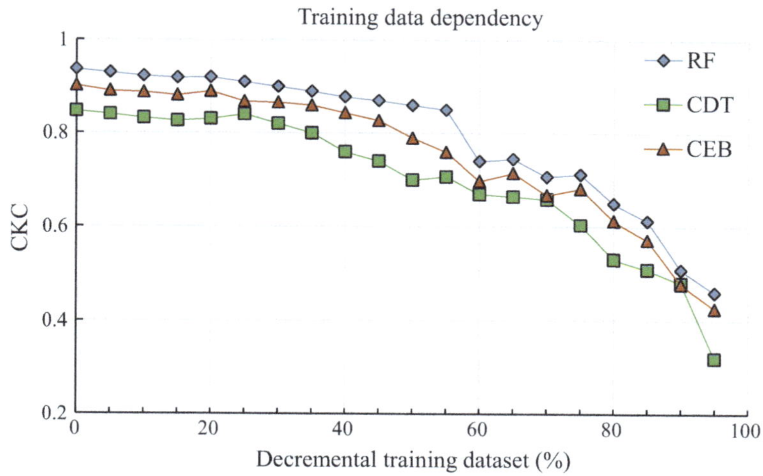

Fig. 4.11 CKC plot for models trained by subsets of the original training data set

mapping accuracy which should be operative in economical and time terms (Foody 1995; Pal and Mather 2003).

Although, data set size is not problematic for the homogeneous area, but for complex areas, large data set is necessary to represent a large number of categorical variability (Foody and Arora 1997; Pal 2005). Some models required a large number of training data and some comparatively less for the same level of performance. However, if an auxiliary data set is not available, then soil training data collection and analysis are the tough job. Therefore, it is an important factor in model selection for DSM.

Figure 4.11 shows CKC plot for models trained by subsets of the original training data in increments of 5%. RF shows the lowest sensitivity to training data set size in the plot. Accuracy reduced by only 0.02 up to 60% data reduction in RF. The abrupt change in accuracy only can be observed for more than 60% data reduction in RF model. CEB also shows stability up to 45% data reduction and a significant decrease in accuracy only can be noted for more than 45% data reduction. However, in CDT, significant decrease can be noted after 25% training data reduction. RF and CEB show more stability than CDT, thus it is concluded that ensemble methods are more stable than single tree model. Furthermore, it is also concluded that less correlation between any two trees in RF should be the reason for more stability than CEB. Therefore, in other words, it can be stated that RF requires a less number of training data in comparison to the other two.

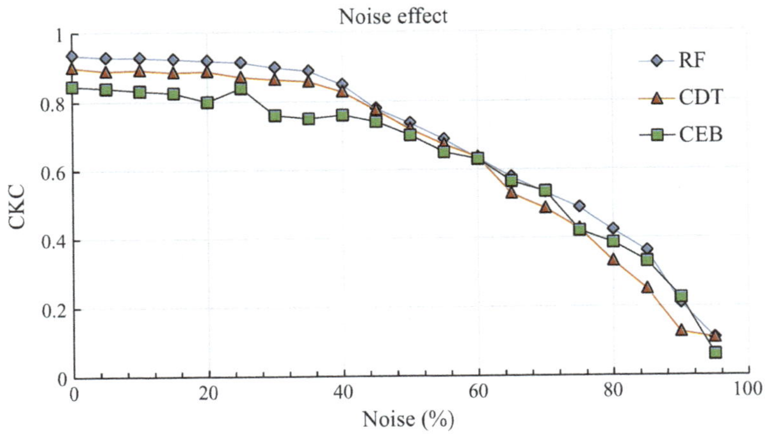

Fig. 4.12 CKC plot against % of noise introduced

4.3.6 *Influence of Noise in the Training Set*

Figure 4.12 shows CKC plot against % of noise introduced in training data by false labelling of the classes. In the plot, each machine learning methods have shown their resistance to noise up to some level. RF and CEB show approximate same behaviour in plot up to 70% noise level.

The slight decrease in CKC value can be observed up to 40% noise level, the only abrupt decrease can be noticed after this for RF and CEB. However, CDT shows more sensitivity to noise in the data set. An abrupt decrease in CKC can be observed after 20% noise level for CDT. Therefore it is concluded that a single decision tree may be prone to noise, but the aggregate of many decision trees reduce the effect of noise giving more accurate results. RF and CEB, both grow the decision trees in the ensemble using bootstrap samples of the data, therefore the effect of noise is less. In other words, RF and CEB both have greater resistance to noise.

4.4 Summary

This chapter provides an effective framework for implementation of non-parametric algorithms for DSM. Effective implementation is in the sense of selection of the optimum value of parameters and selection of suitable variables. Six of seven soil-forming factors, known as scorpan, are selected as environmental covariates to tune RF model. 35 digital layers are prepared using numerous data sources to represent scorpan environmental covariates in the study area. To find the ideal value of parameters *ntree* and *mtry*, CKC is calculated for *mtry* (1–35) and *ntree* (1–1000) values at interval 1. The parameters corresponding to the highest CKC value and lowest *mtry* value are considered as optimum values for RF model.

Furthermore, the importance of variables is also calculated based on a mean decrease in accuracy MDA and MDG score to identify effective indicators. Such estimation is of importance for variable selection in the classification of complex areas where it is mandatory to use large multisource data sets with a large number of variables. Local importance of each class is also measured to know, which, when, and how predictor variable is important to discriminate soil classes. The variables derived from the vegetation indices have shown the strongest influence in the class-separability of the classification scheme. It indicates that 'organism (o)' and its 'spatial context (n)' are an important soil-forming factors for digital soil mapping. Variables related to 'soil property (s)' has also shown the importance and its spatial context is also an important variable. However topographic related variables appeared to have little impact on soil class prediction in the relatively flat surface. Overall ranking in all important measurements indicates that climate variables also contribute significantly.

The behaviour of the RF model is accessed with a change in a number of trees and random split variables, reduction in training data and noise addition in terms of accuracy in OOB and test data set. To understand the effect of parameters *ntree* and *mtry*, the error rate is calculated in OOB and test data set for different *mtry*. The results of this research provide new insights into the performance of RF in the context of soil class mapping. The RF algorithm generates an internal unbiased estimation of the generalization error (OOB error) so it is not necessary to use a test subset independent from the training or resort to cross validation. It is also observed that for any *mtry* value, change in error rate is measurable up to a certain *ntree* value, once the error has converged, the number of random split variables only alter the classifier's accuracy slightly, so the RF can be applied with almost no guidance. However a large number of *mtry* and *ntree* also increase computational time thus, the optimum value is suitable to choose. In outlier measurement, high proximities between the samples of the same class show that almost all the samples are correct.

RF behaviour to a reduction in training data and noise addition is also analyzed based on accuracy. Two other methods, CART decision tree and CART ensemble bagger, are also used to provide the comparative study. RF shows high stability against data set reduction in comparison to the other two. Accuracy reduced by only 0.02 up to 60% data reduction in RF and accuracy decreases more abruptly after this. However, in CDT, significant decrease can be noted after 25% training data reduction. CEB also shows stability up to 45% data reduction. To access the behaviour of models to the false data set, noise in training set is produced by assigning a false class to the training set in 5% increment. The result shows comparatively the greater resistance of RF to noise. Furthermore, comparative performance is also evaluated based on quality assessment measures, i.e. marginal rates, F-measure and JCC, CSI and AC. Performance is scaled between excellent to less than poor performance for each class as well as overall. Each parameter is assigned with a class value corresponding to performance level. The score is calculated on the basis of class value to rank the algorithm. On the basis of class specific measures as well as in multiclass measures, RF scored highest. Total time is also considered as one parameter for performance analysis. Overall RF performed best followed by CEB and CDT.

References

Abdelthaman, S. I., & Onsi, H. M. (1996). Application of remote sensing GIS for soil mapping of El Hasana centeral Sinai, Egypt. In *International Archives of the Photogrammetry, Remote Sensing and Spatial Information Sciences* (Vol. XXXI/B4, pp. 1–6). Vienna: ISPRS.

Boruvka, L., Pavlu, L., Vasat, R., Penizek, V., & Drabek, O. (2008). Delineating acidified soils in the Jizera mountains region using fuzzy classification. In A. E. Hartemink, A. McBratney, & M. L. Mendonça-Santos (Eds.), *Digital soil mapping with limited data* (pp. 303–309). Dordrecht: Springer Netherlands.

Breiman, L. (1996). Bagging predictors. *Machine Learning, 24*(2), 123–140.

Breiman, L. (2001). Random forest. *Machine Learning, 45*(1), 5–32.

Breiman, L., Friedman, J. H., Stone, C. J., & Olshen, R. A. (1984). *Classification and regression trees*. Monterey, California, USA: Wadsworth and Brooks/Cole.

Bui, E. N., & Moran, C. J. (2003). A strategy to fill gaps in soil survey over large spatial extents: An example from the Murray-Darling basin of Australia. *Geoderma, 111*(1–2), 21–44.

Bui, E. N., Loughhead, A., & Corner, R. (1999). Extracting soil-landscape rules from previous soil surveys. *Australian Journal of Soil Research, 37*(3), 495.

Cialella, A., Dubayah, R., Lawrence, W., & Levine, E. (1997). Predicting soil drainage class using remotely sensed and digital elevation data. *Photogrammetric Engineering and Remote Sensing, 63*(2), 171–177.

Cole, N. J., & Boettinger, J. L. (2006). Pedogenic understanding raster classification methodology for mapping soils, powder river basin, Wyoming, USA. In P. Lagacherie, A. B. McBratney, & M. Voltz (Eds.), *Digital soil mapping: An introductory perspective.* Elsevier.

Connolly, J., & Holden, N. M. (2009). Mapping peat soils in Ireland: Updating the derived Irish peat map. *Irish Geography, 42*(3), 343–352.

Dice, L. R. (1945). Measures of the amount of ecologic association between species. *Ecology, 26*(3), 297–302.

Eichhorn, M. P. (2016). What is a natural system? In *Natural systems: The organisation of life* (pp. 341–344). Chichester, UK: Wiley.

Foody, G. M. (1995). Land cover classification by an artificial neural network with ancillary information. *International Journal of Geographical Information Systems, 9*(5), 527–542.

Foody, G. M., & Arora, M. K. (1997). An evaluation of some factors affecting the accuracy of classification by an artificial neural network. *International Journal of Remote Sensing, 18*(4), 799–810.

Genuer, R., Poggi, J.-M., & Tuleau-Malot, C. (2010). Variable selection using random forests. *Pattern Recognition Letters, 31*(14), 2225–2236.

Grimm, R., Behrens, T., Märker, M., & Elsenbeer, H. (2008). Soil organic carbon concentrations and stocks on Barro Colorado island—Digital soil mapping using random forests analysis. *Geoderma, 146*(1–2), 102–113.

Hengl, T., Toomanian, N., Reuter, H. I., & Malakouti, M. J. (2007). Methods to interpolate soil categorical variables from profile observations: Lessons from Iran. *Geoderma, 140*(4), 417–427.

Hilgard, E. W. (1914). *Soils*. London: The Macmillan Company.

Jaccard, P. (1912). The distribution of the flora in the alpine zone. *New Phytologist, 11*(2), 37–50.

Jafari, A., Finke, P. A., Vande Wauw, J., Ayoubi, S., & Khademi, H. (2012). Spatial prediction of USDA- great soil groups in the arid Zarand region, Iran: Comparing logistic regression approaches to predict diagnostic horizons and soil types. *European Journal of Soil Science, 63*(2), 284–298.

Jafari, A., Ayoubi, S., Khademi, H., Finke, P. A., & Toomanian, N. (2013). Selection of a taxonomic level for soil mapping using diversity and map purity indices: A case study from an Iranian arid region. *Geomorphology, 201*, 86–97.

Jenny, H. (1941). *Factors of soil formation. A system of quantitative pedology*. New York: McGraw-Hill Book Company.

Jenny, H., 1941. Factors of soil formation. A system of quantitative pedology. McGraw- Hill Book Company. New York.

Jenny, H. (1980). *The soil resource*. New York, NY: Springer New York.

Joffe, J. S. (1936). *Pedology*. New Brunswick, NJ: Rutgers University Press.

Kempen, B., Brus, D. J., & Heuvelink, G. B. M. (2012). Soil type mapping using the generalised linear geostatistical model: A case study in a Dutch cultivated peatland. *Geoderma, 189–190*, 540–553.

Kim, J., Grunwald, S., Rivero, R. G., & Robbins, R. (2012). Multi-scale modeling of soil series using remote sensing in a wetland ecosystem. *Soil Science Society of America Journal, 76*(6), 2327.

Korolyuk, T. V., & Shcherbenko, H. V. (1994). Compiling soil maps on the basis of remotely-sensed data digital processing: Soil interpretation. *International Journal of Remote Sensing, 15*(7), 1379–1400.

Kovačević, M., Bajat, B., & Gajić, B. (2010). Soil type classification and estimation of soil properties using support vector machines. *Geoderma, 154*(3–4), 340–347.

Labatut, V., & Cherifi, H. (2011). Evaluation of performance measures for classifiers comparison. *Ubiquitous Computing and Communication Journal, 6*, 21–34.

Labatut, V., & Cherifi, H. (2012). Accuracy measures for the comparison of classifiers. In *The 5th International Conference on Information Technology* (p. 11). Amman, Jordan.

Lagacherie, P., Legros, J. P., & Burfough, P. A. (1995). A soil survey procedure using the knowledge of soil pattern established on a previously mapped reference area. *Geoderma, 65*(3–4), 283–301.

Liaw, A., & Wiener, M. (2002). Classification and regression by random forest, *2*(3), 18–22.

Lookingbill, T., & Urban, D. (2004). An empirical approach towards improved spatial estimates of soil moisture for vegetation analysis. *Landscape Ecology, 19*(4), 417–433.

Louppe, G. (2014). Understanding random forests: From theory to practice. University of Liège.

Louppe, G., Wehenkel, L., Sutera, A., & Geurts, P. (2013). Understanding variable importances in forests of randomized trees. *Advances in Neural Information Processing Systems, 26*, 431–439.

Marchetti, A., Piccini, C., Santucci, S., Chiuchiarelli, I., & Francaviglia, R. (2011). Simulation of soil types in Teramo province (Central Italy) with terrain parameters and remote sensing data. *CATENA, 85*(3), 267–273.

McBratney, A. B., Mendonça Santos, M. L., & Minasny, B. (2003). On digital soil mapping. *Geoderma, 117*(1–2), 3–52.

McKenzie, N. J., & Ryan, P. J. (1999). Spatial prediction of soil properties using environmental correlation. *Geoderma, 89*(1–2), 67–94.

Moonjun, R., Farshad, A., Shrestha, D. P., & Vaiphasa, C. (2010). Artificial neural network and decision tree in predictive soil mapping of Hoi Num Rin Sub-Watershed, Thailand. In *Digital soil mapping* (pp. 151–164). Dordrecht: Springer Netherlands.

Moran, C. J., & Bui, E. N. (2002). Spatial data mining for enhanced soil map modelling. *International Journal of Geographical Information Science, 16*(6), 533–549.

Pahlavan Rad, M. R., Toomanian, N., Khormali, F., Brungard, C. W., Komaki, C. B., & Bogaert, P. (2014). Updating soil survey maps using random forest and conditioned latin hypercube sampling in the loess derived soils of northern Iran. *Geoderma, 232–234*, 97–106.

Pal, M. (2005). Random forest classifier for remote sensing classification. *International Journal of Remote Sensing, 26*(1), 217–222.

Pal, M., & Mather, P. M. (2003). An assessment of the effectiveness of decision tree methods for land cover classification. *Remote Sensing of Environment, 86*(4), 554–565.

Pal, M., & Mather, P. M. (2005). Support vector machines for classification in remote sensing. *International Journal of Remote Sensing, 26*(5), 1007–1011.

Panagos, P., Jones, A., Bosco, C., & Senthil Kumar, P. S. (2011). European digital archive on soil maps (EuDASM): Preserving important soil data for public free access. *International Journal of Digital Earth, 4*(5), 434–443.

Pásztor, L., Szabó, J., Bakacsi, Z., Matus, J., & Laborczi, A. (2012). Compilation of 1:50,000 scale digital soil maps for Hungary based on the digital Kreybig soil information system. *Journal of Maps, 8*(3), 215–219.

Poggio, L., Gimona, A., & Brewer, M. J. (2013). Regional scale mapping of soil properties and their uncertainty with a large number of satellite-derived covariates. *Geoderma, 209–210*, 1–14.

Prasad, A. M., Iverson, L. R., & Liaw, A. (2006). Newer classification and regression tree techniques: Bagging and random forests for ecological prediction. *Ecosystems, 9*(2), 181–199.

Ramann, E. (1911). *Bodenkunde*. Berlin: Verlag Julius Springer.

Rodriguez-galiano, V. F., Ghimire, B., Rogan, J., Chica-olmo, M., & Rigol-sanchez, J. P. (2012). An assessment of the effectiveness of a Random Forest classifier for land-cover classification. *ISPRS Journal of Photogrammetry and Remote Sensing, 67*, 93–104.

Scull, P., Franklin, J., & Chadwick, O. A. (2005). The application of classification tree analysis to soil type prediction in a desert landscape. *Ecological Modelling, 181*(1), 1–15.

Shcherbenko, Y. E. V., Asmus, V. V., & Andronikov, V. L. (1992). Digital processing of remote sensing imagery in soil mapping. *Mapping Sciences and Remote Sensing, 29*(3), 248–261.

Shukla, G., Garg, R. D., Srivastava, H. S., & Garg, P. K. (2018). Performance analysis of different predictive models for crop classification across an aridic to ustic area of Indian states. *Geocarto International, 33*(3), 240–259.

Smith, C., Daneshfar, B., Frank, G., Flager, E., & Bulmer, C. (2012). Use of weights of evidence statistics to define inference rules to disaggregate soil survey maps. In B. Minasny, B. Malone, & A. McBratney (Eds.), *Digital soil assessments and beyond* (pp. 215–220). Sydney: CRC Press.

Stum, A. K. (2010). Random forests applied as a soil spatial predictive model in arid Utah. Utah State University.

Taghizadeh-Mehrjardi, R. (2016). Digital mapping of cation exchange capacity using genetic programming and soil depth functions in Baneh region, Iran. *Archives of Agronomy and Soil Science, 62*(1), 109–126.

Taghizadeh-Mehrjardi, R., Sarmadian, F., Minasny, B., Triantafilis, J., & Omid, M. (2014). Digital mapping of soil classes using decision tree and auxiliary data in the Ardakan Region, Iran. *Arid Land Research and Management, 28*(2), 147–168.

The Hindu. (2017). A pioneering step in digital soil mapping. *The Hindu*.

Triantafilis, J., Earl, N., & Gibbs, I. (2012). Digital soil-class mapping across the Edgeroi district using numerical clustering and gamma-ray spectrometry data. In A. B. McBratney (Ed.), *Digital soil assessments and beyond* (pp. 187–191). Sydney: CRC Press.

van Zijl, G., le Roux, P., & Smith, H. (2012). Rapid soil mapping under restrictive conditions in Tete, Mozambique. In B. Minasny, B. Malone, & A. McBratney (Eds.), *Digital soil assessments and beyond* (pp. 335–339). Sydney: CRC Press.

Vaysse, K., & Lagacherie, P. (2015). Evaluating digital soil mapping approaches for mapping GlobalSoilMap soil properties from legacy data in Languedoc-Roussillon (France). *Geoderma Regional, 4*, 20–30.

Zhu, A. X., Hudson, B., Burt, J., Lubich, K., & Simonson, D. (2001). Soil mapping using GIS, expert knowledge, and fuzzy logic. *Soil Science Society of America Journal, 65*(5), 1463.

Chapter 5
Prediction Models for Crop Mapping

Crops are plants that are grown for food. Unless the available productive land which is the main source of human sustenance is protected and used in a scientific way to give better and increased returns there is no hope of human survival. The following chapter focuses on effective implementation of prediction algorithms for crop cover mapping. The various methods that have been, or could be used for crop cover mapping are discussed. The potential indicators that have been, or could be, used in crop prediction modelling are also discussed.

5.1 Introduction

For planning agencies of the country, current and accurate crop-type maps are among the important data set (Brisco et al. 1998; Haboudane et al. 2002; Harris 2003; Metternicht 2003; Arenas-Toledo and Epiphanio 2011; Rilwani and Ikhuoria 2011). The basic tasks include identification of crop type and area estimation. Though to collect this information field survey is an accurate and unbiased method but time and huge human resources requirement make it difficult. Remote sensing methods are better options for large area and seasonal mapping. In recent years, machine learning algorithms have received increasing attention for crop classification. Different machine learning algorithms have been adopted to discriminate crops including SVM (An and Yang 2007; Mathur and Foody 2008; Tan et al. 2011; ChenChi et al. 2012; Sonobe et al. 2014) and ANN (Murthy et al. 2003; Sarkar et al. 2008). Comparative accuracy assessments are also presented by many researchers (Deschamps et al. 2012; Rodriguez-Galiano and Chica-Rivas 2014; Sonobe et al. 2014, 2016; Kumar et al. 2015, 2016). Many traditional machine learning algorithms generate a single model (e.g. a decision tree or neural network). Ensemble methods have been called the most influential development in data mining and machine learning in the past decade

P. K. Garg et al., *Digital Mapping of Soil Landscape Parameters*, Studies in Big Data 72,
https://doi.org/10.1007/978-981-15-3238-2_5

(Weigend et al. 1995; Tumer and Oza 2003). Better predictive performance could be obtained from ensembles than any of the constituent learning algorithms (Dietterich 2000; Rätsch et al. 2001; Punera and Ghosh 2008). Ensembles are useful with all modelling algorithms, but this study focuses on classification trees. Ensemble methods are of different types depending on the way they combine and by the way they train base models. This study involved classic ensemble methods—bagging, ARCing, RF, and gradient boosting for crop prediction.

Large area crop type depends on various variables viz. parent material and soil, phenology, texture, topography, soil moisture, and climate (precipitation, temperature). Different studies either involved topography and/or soil type in association with texture variables. However, climate and phenology have also shown importance in vegetation class prediction (Mellor et al. 2013). Therefore, it is necessary to analyze and identify potential indicators for effective prediction modelling. This chapter is solely dedicated to provide an effective framework for the implementation of non-parametric algorithms for crop cover, identification of sensitive variables, and to analyze comparative performance analysis. To fulfil the above objective, this chapter focuses on the implementation of models for crop-type prediction using climate variables in association with texture, vegetation, soil moisture, phenology, topography, parent material, and soil, followed by comparative performance analysis. Potential of different variables for crop-type classification is also evaluated. In Chap. 3, Developed variables in six covariates categories are representative of soil–landscape relationship in the region under different soil and climate conditions. Therefore, to incorporate dependency of the crop in different variables, all 35 digital layers are used. These layers are used as predictor variables to train predictive models.

Beside the model selection, reference samples (training and testing data) are necessary for model implementation and their assessment. For this study, crop information from agricultural fields are collected in *rabi* season of 2015. A total of seventeen fields are selected in three regions (Bharatpur, Mathura, and Gurgaon region) to collect as representatives of crop pattern over an agricultural land. Random sampling approach is used to collect 50 to 70 sample points per field across the connecting roads.

5.2 Modelling Environment

5.2.1 Modelling Methods

This study involved classic ensemble methods—CEB, ARCing, RF, and TreeNet Gradient Boosting (TNGB) for crop prediction. To evaluate the ensemble models in a crop mapping context, the results achieved by the ensemble models are compared with those produced by a single tree model CDT. CEB, RF, and CDT are described in the previous chapter (Chap. 4, Sect. 4.2.2). A brief description of other ensemble methods is presented in this section.

5.2.1.1 TreeNet Gradient Boosting Machine (TNGBM)

Like other boosting methods, gradient boosting combines weak 'learners' into a single strong learner in an iterative fashion. GBM is an efficient algorithm for converting relatively poor hypotheses into very good hypotheses. Decision trees are used as the weak learner in gradient boosting. GBM involves three elements: (i) a loss function to be optimized, (ii) a weak learner to make predictions, (iii) an additive model to add weak learners to minimize the loss function. Trees are added one at a time, and existing trees in the model are not changed. Gradient boosting is a greedy algorithm and can overfit a training data set quickly. Adaboost was the first successful boosting algorithm (Freund and Schapire 1996, 1997). Then Adaboost was formulated as gradient descent with a special loss function (Breiman 1998). Again in order to handle a variety of loss functions, Adaboost was generalized to gradient boosting (Friedman et al. 2000; Friedman 2001).

If input training set is (x_i, y_i) where, $i = 1, N$; and base learner $h(x; a)$ is L-terminal node regression tree, differentiable loss function is $\varphi(y, F(x))$, then for M iteration, GBM algorithm works in following steps.

Initialize model with a constant value,

$$F_0(x) = \arg\min_{\gamma} \sum_{i=1}^{N} \varphi(y_i, \gamma) \tag{5.1}$$

For $m = 1$ to M, at each iteration m, a regression tree partitions the x-space into L-disjoint regions $\{R_{lm}\}_1^L$ and predicts a separate constant value in each one.

$$h\left(x; \{R_{lm}\}_1^L\right) = \sum_{l=1}^{L} \bar{y}_{lm} 1(x \in R) \tag{5.2}$$

Here $\bar{y}_{lm} = mean_x \in R_{lm}(\tilde{y}_{im})$ is the mean of current pseudo residuals $(\tilde{y}_{im})$ in each region R_{lm}; the parameters of this base learner are the splitting variables and corresponding split points defining the tree, which in turn define the corresponding regions of the partition at the mth iteration. These are induced in a top-down best-fit manner using a least square splitting criterion.

$$\tilde{y}_{im} = -\left[\frac{\partial\varphi(y_i, F(x_i))}{\partial F(x_i)}\right]_{F(x)=F_{m-1}(x)}, i = 1, N \tag{5.3}$$

$$\{R_{lm}\}_1^L = L - \text{terminal node tree}\left(\{\tilde{y}_{im}, x_i\}_1^N\right) \tag{5.4}$$

Because the tree (5.2) predicts a constant value $\overline{y}_{lm}$ within each region R_{lm}, the solution reduces to a simple location estimated based on the loss function.

$$\gamma_{lm} = \arg\min_{\gamma} \sum_{x_i \in R_{im}} \varphi(y_i, F_{m-1}(x_i) + \gamma) \tag{5.5}$$

The current approximation $F_{m-1}(x)$ is then separately updated in each corresponding region.

$$F_m(x) = F_{m-1}(x) + v.\gamma_{lm} 1(x \in R_{lm}) 0 < v \leq 1 \tag{5.6}$$

where parameter v is called the 'learning rate'. Empirically it has been found that using small learning rates (such as $v \leq 0.1$) yields dramatic improvements in the model's generalization ability over gradient boosting without shrinking ($v = 1$). However, it comes at the price of increasing computational time both during training and querying: lower learning rate requires more iterations.

5.2.1.2 Importance Sampled Learning Ensemble (ISLE)

Given a learning task and a relevant data set, we can generate a set of models from a predetermined model family. Bagging bootstraps the training data set (Breiman 1996) and produces a model for each bootstrap sample. RF creates a diverse set of models by randomly selecting a few aspects of the data set while generating each model (Breiman 2001). Adaboost (Freund and Schapire 1996; Friedman et al. 2000; Friedman 2001) and ARCing (Breiman 1997, 1998) iteratively build models by varying case weights (up-weighting cases with large current errors and down-weighting those accurately estimated) and employ the weighted sum of the estimates of the sequence of models. There have been few attempts to unify these ensemble learning methods. One such framework is the ISLE (Friedman and Popescu 2003).

We are to produce a regression model to predict the continuous outcome variable y from p vector of input variables x. Generated models are from a given model family $F = \{f(x, \theta): \theta \in \Theta\}$ indexed by the parameter θ. The final ensemble models considered by the ISLE framework have an additive form:

$$F(x) = w_0 + \sum_{j=1}^{M} w_j f\left(x, \theta_j\right) \tag{5.7}$$

where $\{f(x, \theta j)\}_{j=1}^{M}$ are base learners selected from F. ISLE uses a two-step approach to produce $F(x)$. The first step involves sampling the space of possible models to obtain $\{\theta bj\}_{j=1}^{M}$. The second step proceeds with combining the base learners by choosing weights $\{wj\}_{j=0}^{M}$ in Eq. (6). The pseudo code to produce M models $\left\{f\left(x, \hat{\theta} j\right)\right\}_{j=1}^{M}$ under ISLE (M, v, η) framework is given below:

$F_0(x)$, for $j = 1$ to M,

$$\left(\hat{c}_j, \hat{\theta}_j\right) = \arg\min \sum_{i \in S_j(\eta)} L\big(y_i, F_{j-1}(x_i) + cf(x_i, \theta)\big) \tag{5.8}$$

$$T_j(x) = f\left(x, \widehat{\theta_j}\right) \tag{5.9}$$

$$F_j(x) = F_{j-1}(x) + \nu \hat{c}_j T_j(x) \tag{5.10}$$

$$\text{Return, } \left(\left\{T_j(x)\right\}_{j=1}^{M} \text{ and } F_M(x)\right) \tag{5.11}$$

Here L is a loss function, $S_j(\eta)$ is a subset of the indices $\{1, 2, \ldots, n\}$ chosen by a sampling scheme η, $0 \leq \nu \leq 1$ is a memory parameter. The classic ensemble methods of Bagging, RF, Adaboost, and Gradient Boosting are special cases of ISLE ensemble model generation procedure (Seni and Elder 2010). In Bagging and RF, the weights in 1 are set to predetermined values, i.e. $w_0 = 0$ and $w_j = 1/M$ for $j = 1, 2, \ldots, M$. Boosting calculates these weights in a sequential fashion at each step by having positive memory ν, estimating c_j and takes $F_M(x)$ as the final prediction model.

Friedman and Popescu (2003) recommend learning the weights $\{wj\}_{j=0}^{M}$ using lasso (Seni and Elder 2010). Let $T_j(x_i)_{i=1,m=1}^{n,M}$ be the $n \times M$ matrix of predictions for the n observations by the M models in an ensemble. The weights $\left(w_0, w = \{wm\}_{m=0}^{M}\right)$ are obtained from Eq. 5.12,

$$\hat{w} = \arg\min(y - w_0 1_n - Tw)'(y - w_0 1_n - Tw) + \lambda \sum_{i=1}^{M} |w_m| \tag{5.12}$$

$\lambda > 0$, is the shrinkage operator, larger values of λ decrease the number of models included in the final prediction model. The final ensemble model is given by the below equation,

$$F(x) = w_0 + \sum_{m=1}^{M} w_m T_m(x) \tag{5.13}$$

5.2.1.3 CART Ensembles ARCing (CEA)

Arcing is an acronym for Adaptive Reweighting and Combining (Breiman 1997). Arcing is more successful than bagging in test set error reduction. Arcing algorithm proceeds in sequential steps with a user defined limit of how many steps until termination. Initialize probabilities $\{p(n)\}$ to be equal. At each step, the new training set is selected by sampling from the original training set using probabilities $\{p(n)\}$.

After the classifier based on this resampled training set is constructed, the $\{p(n)\}$ are updated depending on the misclassifications up to the present step. On termination, the classifiers are combined using weighted (ARC-fs) or un-weighted (ARC-x4) voting. The arc-fs algorithm based on a boosting theorem is given in Freund and Schapire (1997). ARC-x4 is an ad hoc invention. Both arcing algorithms are given below.

ARC-fs ARCing:

1. At the kth step, using the current probabilities $\{p^{(k)}(n)\}$, a sample with replacement from T to get the training set $T^{(k)}$ and construct classifier C_k using $T^{(k)}$.
2. Run T down the classifier C_k and let $d(n) = 1$ if the nth case is classified incorrectly; otherwise zero.
3. Define

$$\varepsilon_k = \sum_n p^{(k)}(n)d(n),\ \beta_k = (1 - \varepsilon_k)/\varepsilon_k \tag{5.14}$$

And the updated $k + 1$ step probabilities by the given equation

$$p^{(k+1)}(n) = \frac{p^{(k)}(n)\beta_k^{d(n)}}{\sum p^{(k)}(n)\beta_k^{d(n)}} \tag{5.15}$$

After K steps, the $C_1, \ldots, C_k$ are combined using weighted voting with C_k having weight $\log(\beta_k)$. Breiman (1998) made two minor revisions to this algorithm. If ε_k becomes greater than 1/2, then the voting weights β_k go negative. They found that setting all $\{p^{(k)}(n)\}$ equal and restarting gives good results.

ARC-x4 ARCing:

1. At the kth step, using the current probabilities $\{p^{(k)}(n)\}$, a sample with replacement from T to get the training set $T^{(k)}$ and construct classifier C_k using $T^{(k)}$.
2. Run T down the classifier C_k and let $m(n)$ be the number of misclassifications of the nth case by $C_1, \ldots, C_k$.
3. The updated $k + 1$ step probabilities are defined by

$$p(n) = \frac{\left(1 + m(n)^4\right)}{\sum\left(1 + m(n)^4\right)} \tag{5.16}$$

After K steps, the $C_1, \ldots, C_k$ are combined by un-weighted voting. ARC-x4 has a slight edge on the smaller data sets, while AEC-fs does a little better on the larger ones.

5.2.2 Construction and Implementation

Matlab is used for model construction and implementation of RF predictive model. The randomForest-matlab (@misc) package is used to build the RF model. The code is based on R source by Liaw and Wiener (2002) of RF algorithm, which is originally developed by Leo Breiman and Adele Cutler in FORTRAN (Breiman 2001; Liaw and Wiener 2002).

In the previous study (Chap. 4, Sect. 4.3.1), it was suggested that (i) RF should be optimized to find the optimum *mtry* and *ntree* value, (ii) the OOB estimation is as accurate as the estimation based on the test data set. Therefore, to find the ideal value of parameters *ntree* and *mtry*, out of bag (OOB) error rate is calculated (instead of CKC in test data set, as in case of soil study) for *mtry* between D/10 (D = number of predictor variables) and $\sqrt{D}$ with 1000 trees. OOB error rate plot showed the lowest OOB error rate for *mtry* value 5 and stabilization of overall error at 650 trees (Fig. 5.1). Therefore, 5 is used for parameter *mtry* and 650 for parameter *ntree* in the model. It means, model will produce forest of 650 trees and at every node, it will choose 5 predictor variables at random.

The Salford Predictive Modeller (SPM) of Salford System is used for the implementation of other models. It included CART (Classification and Regression Tree) methods. TreeNet and ISLE, CDT, CEB, and CEA are used in CART (Breiman et al. 1984; Steinberg and Colla 1995). TreeNet data mining tool is used to implement gradient boosting approach to achieve the benefit of boosting without the drawback of a tendency to be misled by bad data. It consists of several hundred small trees.

The score is developed separately for each class via class specific expansions, and the scores are converted into a set of probabilities of class membership. Similar to RF, 650 trees are selected to build with 5 as a maximum number of nodes per tree.

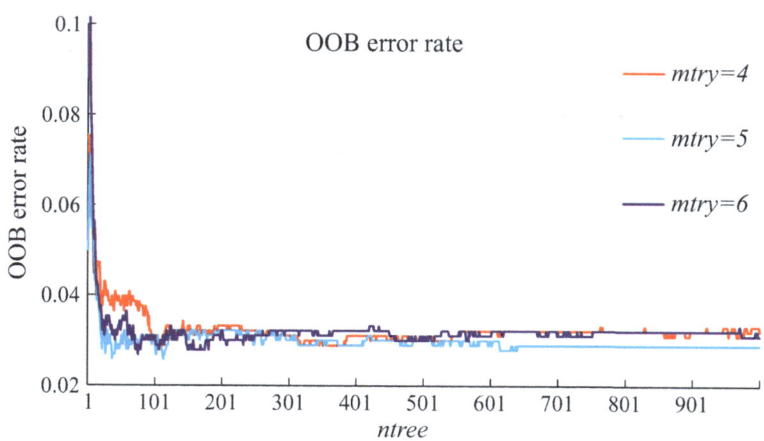

Fig. 5.1 OOB error rate convergence for *mtry* values 4, 5, and 6 with 1000 trees

Equal weight is assigned to each class. Independent data set (test data) of training samples are used to evaluate models. In addition to test data subset of bootstrap data set, i.e. OOB is also used to evaluate RF model. All 35 variables are selected for training as well as for testing. The total study area is divided into four equal parts to avoid out of memory problem in Dell PC of 8 GB RAM, 500 GB hard disk and processor: Intel (R) Core (TM) I 7-2600 CPU @ 3.40 GHz. Models are implemented independently to predict classes in each part and mosaicked together into a single image (Figs. 5.2 and 5.3).

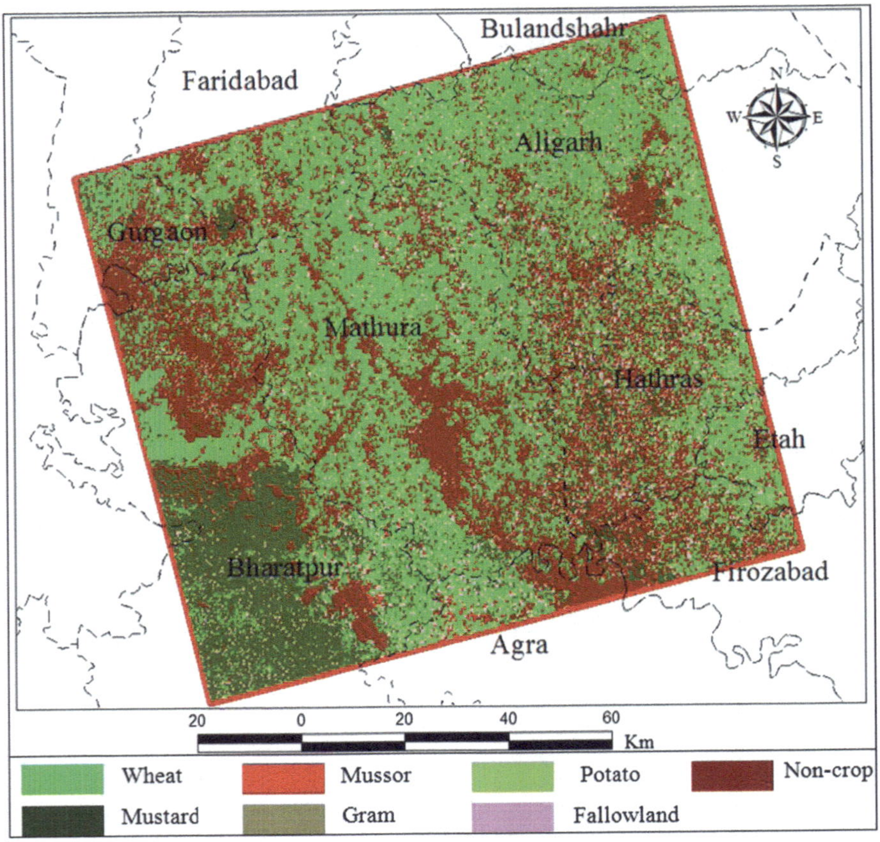

Fig. 5.2 Crop and non-crop map of the study area using RF model

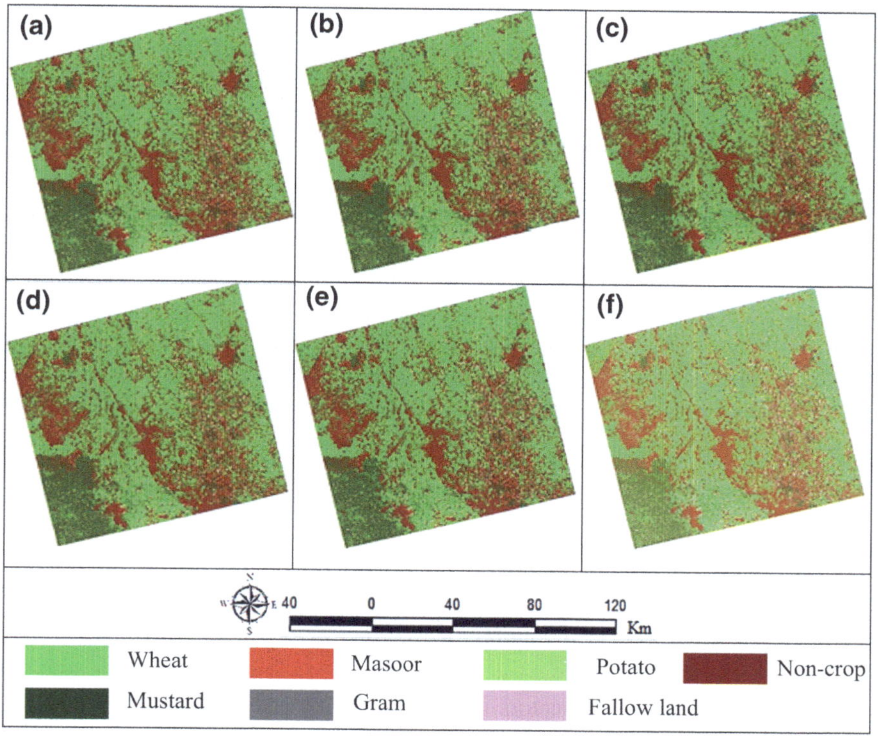

Fig. 5.3 Crop and non-crop maps of the study area using different models; **a** RF; **b** TNGB; **c** CEB; **d** CEA; **e** CDT; and **f** ISLE

5.2.3 Performance Evaluation

5.2.3.1 Class-Separability Measure

To determine the spectral separability among crop types, the Jeffries Matusita (JM) distance method (Richards 2013) have been used. The JM measurement is obtained from Bhattacharya Distance (BD). Both JM and BD are devised to indicate how well a selected spectral class pair is statistically separate. For class 'a' and 'b', it can be expressed as

$$\mathrm{BD} = \frac{1}{8}(\mu_a - \mu_b)^T \left(\frac{C_a + C_b}{2}\right)^{-1} (\mu_a - \mu_b) + \frac{1}{2}\ln\left[\frac{\frac{1}{2}|C_a + C_b|}{\sqrt{|C_a| \times |C_b|}}\right] \tag{5.17}$$

$$\mathrm{JM} = \sqrt{2\left(1 - \mathrm{e}^{-\mathrm{BD}}\right)} \tag{5.18}$$

where μ_a and μ_b are the mean of two classes and C_a and C_b are the covariance matrices for the same classes, and T denotes the transpose of a vector. JM distance has an upper boundary of 1.41(2) and lower boundary of 0. JM distance value <1.0 indicates poor separability and 0 indicates totally inseparable classes.

5.2.3.2 Assessment Parameters

This study also adopted popular accuracy measures as suggested by Labatut and Cherif (2011) and classifiers are evaluated using accuracy measures; (i) Overall Success Rates (OSR), (ii) marginal rates, (iii) F-measure and Jaccard's coefficient of community (Jaccard 1912), (iv) Classification Success Index (CSI), and (v) Agreement Coefficients (AC) (Labatut and Cherifi 2011, 2012). The grading system is adopted to assign a class value for each measurement and the performance of models is evaluated by the ranking method, as described in Sect. 4.2.4 Performance evaluation (Chap 4).

5.3 Results and Discussion

5.3.1 Identification of Effective Indicators

Identification of effective indicators is based on MDA and MDG score (Fig. 5.4a, b). The MDA is based on the hypothesis that if the variable is not important, then permuting the value of that variable will not degrade prediction accuracy (Genuer et al. 2010). The MDG ranks variables according to how cleanly the variable separated the classes when selected at a node. In the previous chapter (Chap. 4), more details of these measures are described in Sect. 4.2.2 Soil spatial prediction models (Sect. 4.2.2.1 Random forest classifier). The local importance of each class is also measured (Fig. 5.5) to know, which, when, and how predictor variable is important to discriminate crop class within crop classes and to crop from non-crop classes.

Landsat-8 band 5 (near infrared band) is the most important predictor variable based on MDA and MDG scores. Other Landsat-8 spectral bands: B2, B3, B4 are also in the top ten most important predictor variables in both scores. Out of six normalized difference band ratios variables, three variables: B5 – B7/B5 + B7, B4 – B2/B4 + B2, and B5 – B6/B5 + B6 are among the top ten most important variables when estimating using MDA as well as in MDG. This shows the importance of soil information in class discrimination. B-6 and B-7 spectral bands of Landsat-8 also scored well except for *masoor* and *gram* crop classes, where B-7 score is low in local importance plots. The B-5 band which is the most important variable based on MDA and MDG scores also recorded the lowest value for *masoor* and *gram* in local importance plot. This variation indicates the importance of spectral responses in the model.

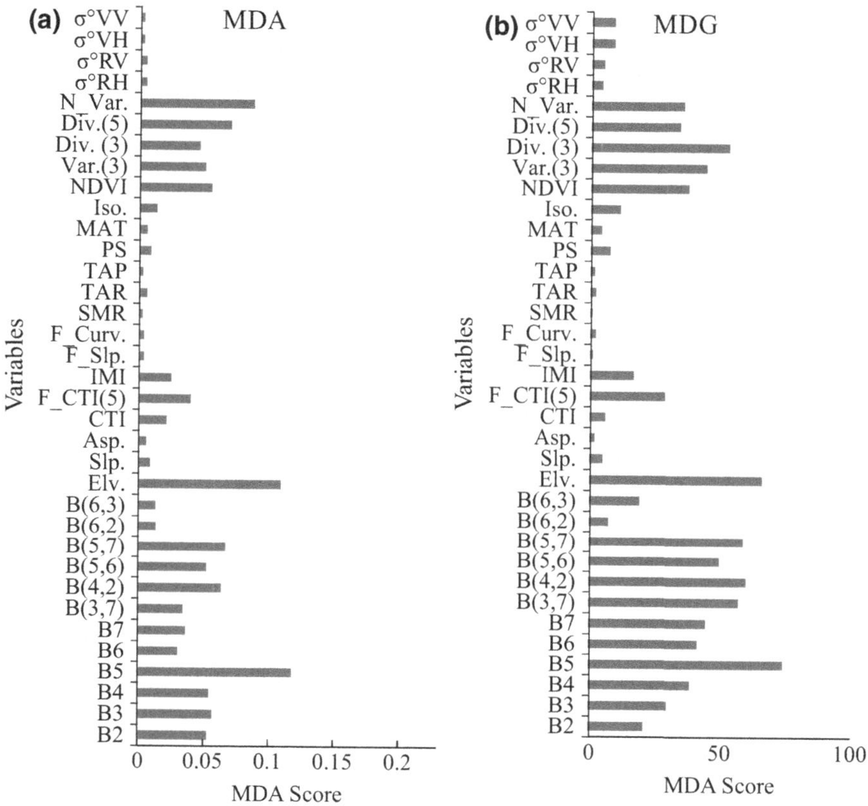

Fig. 5.4 Potentiality of variables based on **a** MDA and **b** MDG score

The elevation is the second most important predictor variable in both scores. From the local importance of each class, it can be concluded that a high score of elevation variable in MDA is due to its importance to predict crop classes. Predictor variable elevation is not even in the top ten important variables in the local importance plot for non-crop.

However, DEM derived topographic variables; slope and aspect are not among the top ten variables in the MDA and MDG measurement. Local importance for each class is also not indicating the importance of these variables for class prediction. Therefore, slope and aspect appeared to have little impact on class prediction in the relatively flat surface.

Other topographic variables associated with soil attributes (CTI, IMI) scored well in MDG score. Predictor variable filtered CTI is in 8th position when estimating using the MDG score. It indicates that the potential of CTI and IMI is useful in class prediction in the model. However, these variables are not in the top ten important variables, when estimated using MDA score or class wise score (local importance).

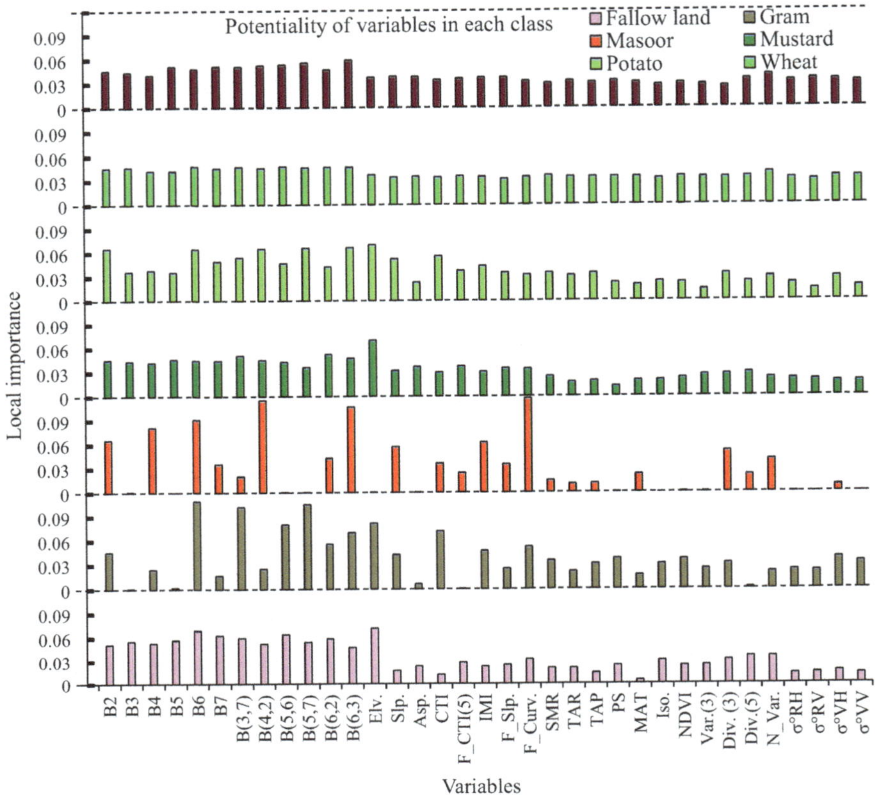

Fig. 5.5 Local importance plot showing potentiality of variables in each class

NDVI variance is the third most important predictor variable based on MDA score and fourth in MDG score, indicating the importance of the phenological temporal variance in NDVI to predict classes. For crop fields, the standard deviation is large in comparison to other land covers. However, other vegetated areas, e.g. forest, does not show much variation in NDVI value throughout the year, i.e. low standard deviation. NDVI and NDVI derived variables are also among the top ten most important predictor variables in both the scores. Therefore, it indicated that variables associated to vegetation and their texture measurements are useful in the prediction model. However, in local importance plot, spectral bands and their normalized difference ratios dominate over other variables. *Masoor* and *gram* reflected different behaviours in terms of variable importance.

Climate variables have not shown their importance as much as expected for class prediction in any of the importance measurements. It was expected to achieve a good score for climate parameter because these parameters govern differences in albedo, crop height, aerodynamic properties, and leaf and stomata properties of different crops (Wang et al. 2016). Similarly, the backscattering responses in two polarization modes have also scored lower values in MDA and MDG as well as have not shown importance in local importance plot.

Three types of variables are involved in this study, i.e. spectral (Landsat-8 and SAR), topographic (DEM derived), climate. Topographic and spectral variables (except SAR) have major importance in class prediction over climate variables. These variables are used separately to produce a forest of 650 trees with *mtry* value 5 and OOB error rate has been calculated to test the comparative importance of topographic, spectral (only Landsat-8 derived), and climate variables. OOB error rate for the spectral only is 0.05, for topographic only is 0.065, and for climate only is 0.076. This indicates the importance of spectral variables over topographic and climate variables. However, in the final RF model, all these three types of variables have a contribution.

In the previous chapter (Chap. 4), it is concluded that RF is less sensitive to training data set size and has a greater resistance to noise. For crop-type mapping in a large area, it is difficult in both economic and time terms to collect large training data set. However, for the homogeneous area, training data size is not problematic but for complex areas, with a large number of categories, it is necessary to have training set be as large as possible to represent all the variabilities present in a category and selected variables should truly represent variability in classes.

Selection of variables at each split in RF also increases the probability of choosing an important variable at a split. Therefore RF has an advantage over other prediction methods for large area crop mapping. However in this study, RF is implemented for crop mapping in *rabi* season and different possible variables are also tested but the model should be also tested for *kharif* crops (*rice*, *maize*, *sorghum*, *bajra*, *finger millet*, *arhar*, *soyabean*, *groundnut*, *cotton*) with same variables.

5.3.2 Class-Separability Analysis

The separability analysis using JM method has shown a better separation between almost all classes. Especially in the case of non-crop class, the separability is found better than the crop classes due to its unique and high spectral reflectance. Its values >1.7 demonstrate that the classes are well separated. The separability analysis using JM method is given in Table 5.1.

5.3.3 Comparative Performance Analysis

5.3.3.1 RF Classifier

In test data set for *gram* and *masoor* class, PPV and NPV $\approx$1, indicated perfect prediction (i.e. classifier prediction is 100% reliable). All marginal rates scored excellent measures value. All other class specific measures are a function of TPR and PPV. For

Table 5.1 Separability analysis between different crop and non-crop classes using JM distance method

	Fallow land	Gram	Masoor	Mustard	Potato	Wheat	Non-crop
Fallow land	0	1.89	1.94	2	1.92	1.87	1.98
Gram	1.89	0	1.78	1.93	1.94	2	2
Masoor	1.94	1.78	0	1.92	1.87	1.86	2
Mustard	2	1.93	1.92	0	2	1.89	1.97
Potato	1.92	1.94	1.87	2	0	1.95	1.93
Wheat	1.87	2	1.86	1.89	1.95	0	1.93
Non-crop	1.98	2	2	1.97	1.93	1.93	0

fallow land, *gram*, *masoor*, *potato*, and *non-crop* class, all class specific measures values are under excellent value range. In the case of *fallow land,* TPR = PPV and TNR = NPV, it showed the same level of prediction reliability and performance of classifier for this class. *Mustard* and *wheat* class also scored excellent values for these measures except JCC and ICSI, which scored slightly lower values due to slightly lower TPR and PPV value. All multiclass measures also scored excellent values. However, in OOB data set, poor value of PPV of *masoor* class (0.6556, i.e. only 65.56% predicted *masoor* class belongs to *masoor)* showed that classification prediction is less reliable for this class but other measures values are high, 99.89% of predicted non-masoor class actually do not belongs to *masoor* (NPV), 96.84% non-masoor class do not put in predicted *masoor* class (TNR), and 98.33% *masoor* class are correctly classified (TPR). ICSI, JCC, and F-measures are a function of PPV and TPR and low value of PPV also reflected in these measures value for *masoor* class. However, for other classes, it showed its good reliability with fair to the high value of predictive rates. For fallow land, PPV and NPV $\approx$1, indicated perfect prediction. High TPR and TNR values indicated classifier is good at classifying instances belonging to a specific true class, as much for instances belonging to other classes. High value of other class specific measures as well as all multiclass measures also (CKC: 0.9167; OSR: 0.9356; CSI: 0.8359) indicated good performance of classifier. Test performance is better than on OOB data (Table 5.2).

5.3.3.2 CDT

The marginal rates scored good to excellent measures values for all classes. Although PPV value (0.8235, i.e. only 82.35% predicted *potato* class belongs to *potato* class) of *potato* is good but in comparison to NPV (0.998; i.e. 99.8% predicted non-potato class do not belong to *potato* class), low value indicated that classifier prediction is more reliable for other classes than *potato*. Similarly, for *gram* class, significant difference between PPV (0.8966) and NPV (0.9980) also indicated that *gram* class prediction is not as much as reliable as for other classes. For all other classes, PPV and NPV difference <0.04, indicated the same reliability for predicted individual class and for

Table 5.2 Performance measures of classifiers RF, CDT, and CEB

		Fallow land	Gram	Masoor	Mustard	Potato	Wheat	Non-crop	Score
RF-test	TPR	0.9787^5	0.9630^5	0.9688^5	0.9320^5	0.9333^5	0.9394^5	0.98^5	35
	TNR	0.9979^5	1^5	1^5	0.9857^5	1^5	0.9882^5	0.9721^5	35
	PPV	0.9787^5	1^5	1^5	0.9411^5	1^5	0.9490^5	0.9560^5	35
	NPV	0.9979^5	0.9980^5	0.9980^5	0.9833^5	0.9980^5	0.9859^5	0.9874^5	35
	F-meas.	0.9787^5	0.9811^5	0.9841^5	0.9365^5	0.9655^5	0.9444^5	0.9679^5	35
	JCC	0.9583^5	0.9630^5	0.9688^5	0.8807^4	0.9333^5	0.8942^4	0.9378^5	33
	ICSI	0.9574^5	0.9630^5	0.96875^5	0.8732^4	0.9333^5	0.8883^4	0.9360^5	33
	Score	35	35	35	33	35	33	35	
	CKC: 0.9472^5; OSR: 0.9598^5; CSI: 0.9314^5; Total time: $00{:}0{:}49^5$								
RF-OOB	TPR	0.99^5	0.96^5	0.9833^5	0.89^4	0.9667^5	0.945^5	0.9275^5	34
	TNR	1^5	0.9848^5	0.9684^5	0.9869^5	0.9950^5	0.9952^5	0.9984^5	35
	PPV	1^5	0.7619^3	0.6556^2	0.9418^5	0.8529^4	0.9792^5	0.9973^5	29
	NPV	0.9989^5	0.9980^5	0.9989^5	0.9741^5	0.999^5	0.9870^5	0.9566^5	35
	F-meas.	0.9950^5	0.8496^4	0.7867^3	0.9152^5	0.9063^5	0.9618^5	0.9611^5	32
	JCC	0.99^5	0.7385^3	0.6484^2	0.8436^4	0.8286^4	0.9264^5	0.9252^5	28
	ICSI	0.99^5	0.7219^3	0.6389^2	0.8318^4	0.8198^4	0.9242^5	0.9248^5	28
	Score	35	28	24	32	32	35	35	
	CKC: 0.9167^5; OSR: 0.9356^5; CSI: 0.8359^4; Total time: $00{:}00{:}34^5$								
CDT	TPR	0.9787^5	0.9630^5	0.9688^5	0.8878^4	0.9333^5	0.9192^5	0.9698^5	34
	TNR	0.9979^5	0.9939^5	1^5	0.9832^5	0.9940^5	0.9856^5	0.9717^5	35
	PPV	0.9787^5	0.8966^4	1^5	0.9255^5	0.8235^4	0.9381^5	0.9554^5	33

(continued)

Table 5.2 (continued)

		Fallow land	Gram	Masoor	Mustard	Potato	Wheat	Non-crop	Score
	NPV	0.9979^5	0.9980^5	0.9979^5	0.9740^5	0.998^5	0.9810^5	0.981^5	35
	F-meas.	0.9787^5	0.9286^5	0.9841^5	0.9062^5	0.875^4	0.9286^5	0.9626^5	34
	JCC	0.9583^5	0.8667^4	0.9688^5	0.8286^4	0.7778^3	0.8667^4	0.9279^5	30
	ICSI	0.9574^5	0.8595^4	0.9688^5	0.8132^4	0.7569^3	0.8573^4	0.9252^5	30
	Score	35	32	35	32	29	33	35	
	CKC: 0.9265^5; OSR: 0.9439^5; CSI: 0.8769^4; Total time: $00{:}00{:}40^5$								
CEB	TPR	0.9787^5	0.9630^5	0.9688^5	0.8811^5	0.9333^5	0.9293^5	0.98^5	35
	TNR	0.9979^5	0.9960^5	1^5	0.9833^5	0.9980^5	0.9834^5	0.9720^5	35
	PPV	0.9787^5	0.9286^5	1^5	0.9271^5	0.9333^5	0.9293^5	0.9561^5	35
	NPV	0.9979^5	0.9980^5	0.9980^5	0.9717^5	0.9980^5	0.9834^5	0.9873^5	35
	F-meas.	0.9787^5	0.9454^5	0.9841^5	0.9036^5	0.9333^5	0.9293^5	0.9679^5	35
	JCC	0.9583^5	0.8966^4	0.969^5	0.8241^4	0.875^4	0.8679^4	0.9378^5	31
	ICSI	0.9574^5	0.8915^4	0.9688^5	0.8082^4	0.8667^4	0.8585^4	0.9360^5	31
	Score	35	33	35	33	33	33	35	
	CKC: 0.9320^5; OSR: 0.9481^5; CSI: 0.8982^4; Total time: $00{:}04{:}40^1$								

other classes. For the *mustard,* significant difference between TNR (0.9832) and TPR (0.8878) indicated that performance of the classifier is better for other classes than instances belonging to *mustard.* For all classes, TNR values are greater than TPR and it also showed the same behaviour of the classifier as for *mustard.* However, TPR value is good (0.9333, 93.33% prediction is right), the effect of relatively low PPV values (0.8235) can observe in other class specific measures for *potato* class. For *masoor* class, PPV and NPV $\approx$1, indicated perfect prediction. F-measure (0.875, i.e. 87.50% overlap between predicted and true data set) scored good but fair values of JCC (0.7778) and ICSI (0.7569) for *potato* class. Multiclass measures CKC (0.9265), OSR (0.9439), and CSI (0.8769) values are also good (Table 5.2).

5.3.3.3 CE B/A

For *fallow land, masoor,* and *potato* classes, both ensembles methods have the same measures values. TPR, PPV, and TNR, NPV values are the same for *fallow land* and *potato* class, it's indicated that both methods are the same in performance and as well as in reliability for prediction for these classes. All marginal rates and F-measures scored excellent measures value in both methods for all classes, except TPR value of a *mustard* class that lies under the good level of measures. Overall all class specific measures, as well as multiclass measures scored good to excellent values in both ensembles methods. However, JCC and ICSI measures values are comparatively low for *mustard* class. For *masoor* class, PPV and NPV $\approx$1, indicated perfect prediction in both methods. However, ensembles ARCing method showed perfect prediction also for *gram* class. All multiclass measures in ARCing scored slightly better than bagger ensembles (Tables 5.2 and 5.3).

5.3.3.4 ISLE

ISLE showed the lowest reliability for *gram* and *potato* classes prediction with PPV measure values 0.5777 (only 57.77% of the predicted *gram* class actually belong to *gram*) and 0.5833 (only 57.77% of the predicted *potato* class actually belong to *potato*), respectively. However, these classes scored excellent NPV values. It indicated that classifier predictions are reliable for classes other than *gram* and *potato* but it fails on the prediction of these classes. Less than poor value PPV influenced JCC and ICSI measures value associated to these classes (*gram*- JCC: 0.5652, ICSI: 0.5407; *potato*- JCC: 0.561, ICSI: 0.5167) and degraded overall CSI (0.7910) measure value of method. For *masoor* class, PPV and NPV $\approx$1, indicated perfect prediction. Apart from that class specific measures as well as multiclass measures scored fair to excellent values (Table 5.3).

Table 5.3 Performance measures of classifier CEB/A, ISLE, and TNGBM

		Fallow land	Gram	Masoor	Mustard	Potato	Wheat	Non-crop	Score
CEA	TPR	0.9787^{5}	0.9629^{5}	0.9688^{5}	0.8932^{4}	0.9333^{5}	0.9495^{5}	0.985^{5}	34
	TNR	0.9979^{5}	1^{5}	1^{5}	0.9880^{5}	0.9980^{5}	0.9835^{5}	0.9721^{5}	35
	PPV	0.9787^{5}	1^{5}	1^{5}	0.9484^{5}	0.9333^{5}	0.9307^{5}	0.9563^{5}	35
	NPV	0.9979^{5}	0.99799^{5}	0.9980^{5}	0.9742^{5}	0.9980^{5}	0.9882^{5}	0.9905^{5}	35
	F-measure	0.9787^{5}	0.9811^{5}	0.9841^{5}	0.92^{5}	0.9333^{5}	0.94^{5}	0.9704^{5}	35
	JCC	0.9583^{5}	0.9630^{5}	0.9688^{5}	0.8519^{4}	0.875^{4}	0.8868^{4}	0.9426^{5}	32
	ICSI	0.9574^{5}	0.9630^{5}	0.9688^{5}	0.8417^{4}	0.8667^{4}	0.8802^{4}	0.9413^{5}	32
	Score	35	35	35	32	33	33	35	
	CKC: 0.9423^{5}; OSR: 0.9560^{5}; CSI: 0.9170^{5}; Total time: $00{:}04{:}40^{1}$								
ISLE	TPR	0.9574^{5}	0.9630^{5}	0.9688^{5}	0.8571^{4}	0.9333^{5}	0.9495^{5}	0.90^{5}	34
	TNR	0.9979^{5}	0.9613^{5}	1^{5}	0.9905^{5}	0.9801^{5}	0.9832^{5}	0.990^{5}	35
	PPV	0.9782^{5}	0.5777^{1}	1^{5}	0.9545^{5}	0.5833^{1}	0.9306^{5}	0.9836^{5}	27
	NPV	0.9957^{5}	0.9978^{5}	0.9979^{5}	0.9674^{5}	0.9979^{5}	0.9880^{5}	0.9402^{5}	35
	F-measure	0.9677^{5}	0.7222^{3}	0.9841^{5}	0.9032^{5}	0.7179^{3}	0.94^{5}	0.9399^{5}	31
	JCC	0.9375^{5}	0.5652^{1}	0.9687^{5}	0.8235^{4}	0.56^{1}	0.8867^{4}	0.8867^{4}	24
	ICSI	0.9357^{5}	0.5407^{1}	0.9687^{5}	0.8117^{4}	0.5167^{1}	0.8802^{4}	0.8836^{4}	24
	Score	35	21	35	32	21	33	33	
	CKC: 0.8906^{4}; OSR: 0.9151^{5}; CSI: 0.7910^{3}; Total time: $00.02{:}25^{2}$								
TNGBM	TPR	0.9787^{5}	0.9629^{5}	0.9687^{5}	0.9029^{5}	0.9333^{5}	0.9394^{5}	0.9799^{5}	35
	TNR	1^{5}	0.9960^{5}	1^{5}	0.9880^{5}	1^{5}	0.9810^{5}	0.9721^{5}	35
	PPV	1^{5}	0.9285^{5}	1^{5}	0.9490^{5}	1^{5}	0.9208^{5}	0.9559^{5}	35

(continued)

Table 5.3 (continued)

	Fallow land	Gram	Masoor	Mustard	Potato	Wheat	Non-crop	Score
NPV	0.9979^{5}	0.9980^{5}	0.9980^{5}	0.9764^{5}	0.9980^{5}	0.9857^{5}	0.9874^{5}	35
F-measure	0.9892^{5}	0.9454^{5}	0.9841^{5}	0.9253^{5}	0.9655^{5}	0.93^{5}	0.9677^{5}	35
JCC	0.9787^{5}	0.8966^{4}	0.9688^{5}	0.8611^{4}	0.9333^{5}	0.8692^{4}	0.9375^{5}	32
ICSI	0.9787^{5}	0.8915^{4}	0.9688^{5}	0.8519^{4}	0.9333^{5}	0.8602^{4}	0.9358^{5}	32
Score	35	33	35	33	35	33	35	
CKC: 0.9397^{5}; OSR: 0.9540^{5}; CSI: 0.9172^{5}; Total time: $00{:}00{:}34^{1}$								

Table 5.4 Performance scale for model evaluation

Accuracy measures	Class value	Score	Classifier performance
0.90–1.0	5	31.5–35	Excellent
0.80–0.90	4	28–31.5	Good
0.70–0.80	3	24.5–28	Fair
0.60–0.70	2	21–24.5	Poor
<0.60	1	17.5–21	Less than poor

5.3.3.5 TNGBM

For *fallow land*, *masoor*, and *gram* classes, PPV and NPV $\approx$1, indicated perfect prediction for these classes. Excellent values of marginal rates for all classes showed that classifier is excellent in classifying instances belonging to these classes and also excellent prediction reliability. All other class specific measures are functions of positive rates hence the high value of these measures is obvious. Multiclass measures also scored excellent values (Table 5.3).

To rank the methods, a comparative analysis is performed using score (Table 5.4). Each measure value is assigned a class value ranging from 1 to 5. Class value is corresponding to different ranges of accuracy measures. High accuracy range (90–100%) is corresponding to the high class value (5). For each accuracy measures corresponding to a particular class and for particular measure corresponding to each class, the score is calculated from assigned class values. Total time (training + testing time) is also included to evaluate performance. To assign class value to time, lowest time (TL) is assigned a highest class value (C = 5) and the fraction is calculated for other time value (f = T/TL) to calculate the class value (C/f). The overall performance of methods is evaluated based on total score.

Figure 5.6 shows the performance of methods on the basis of class specific measures. All methods equally performed for *fallow land* with the highest score (35). ISLE shows the lowest performance for *potato* and *gram* classes. For *masoor* and

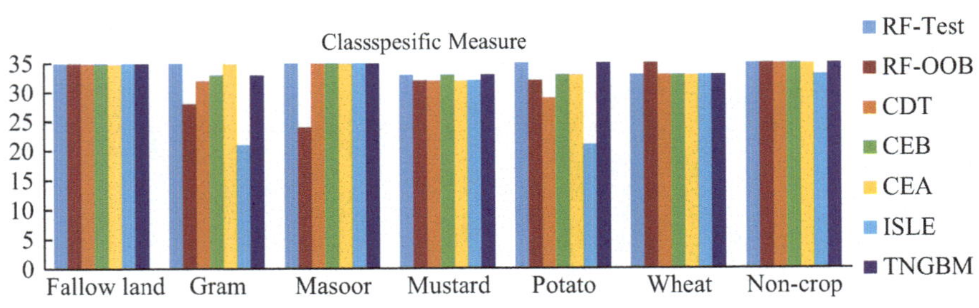

Fig. 5.6 Performance of methods on the basis of class specific measures

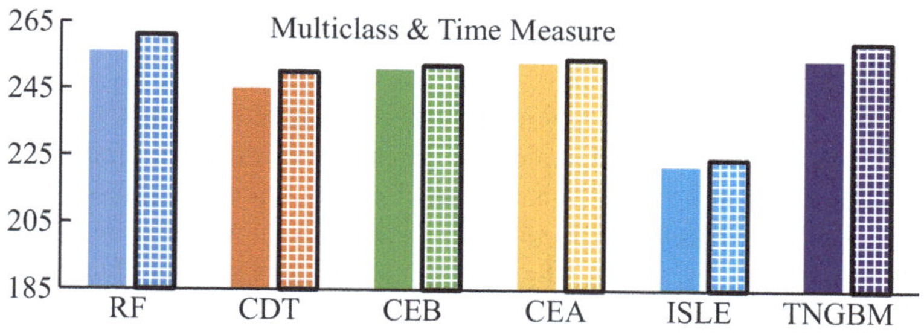

Fig. 5.7 Performance of methods on the basis of multiclass measures and total time (dark boundary)

wheat class, all methods scored equal value (*masoor*: 35, *wheat*: 33) except for OOB data set in RF (*masoor*: 24, *wheat*: 35). Except ISLE (33), all methods get the highest score for *non-crop* (35). For *gram* class, different methods show different levels of performance while for *mustard* nearly equal performance.

Multiclass measures score is also used to evaluate methods performance (Fig. 5.7). It is the sum of the class value assigned to multiclass measures. OOB data set in RF is not considered in the evaluation because only test data is similar in all methods. RF, CEA, and TNGBM are top scorer followed by CDT, and CEB, while ISLE gets the lowest score. Figure 5.7 (das bar) shows performance on the basis of total time taken in training and testing. RF, CDT, and TNGBM get the highest score (lowest time) followed by ISLE. CEB and CEA scored the lowest value (i.e. highest time).

The overall performance of methods is evaluated using the sum of scores of class specific and multiclass measures. The further time factor is also included to compare methods performance. Figure 5.8 shows overall performance without and with time factor (das bar), respectively. In both cases, RF is the best performer followed by TNGBM. CEA, CEB, and CDT get rank 3rd, 4th, and 5th, respectively. ISLE performed worst.

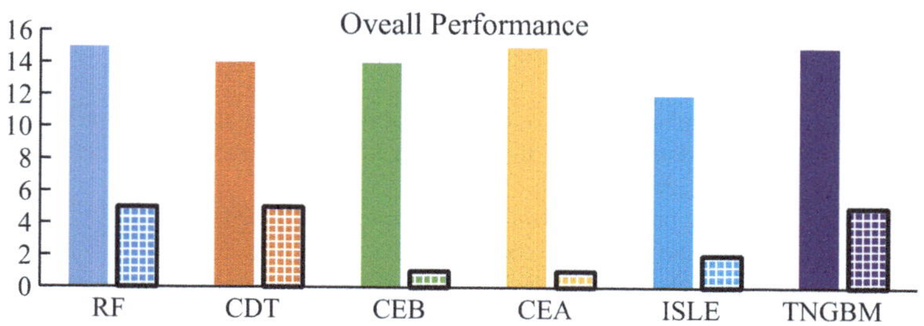

Fig. 5.8 Overall performance of methods without time and with time factor (dark boundary)

5.4 Summary

This study focused on to effectively implement the predictive models for large area crop prediction and to check classification capability of different predictive models. To effectively tune to discriminate crop class from *non-crop* classes and sub-classification of crop class, climate variables in association with texture, vegetation, soil moisture, phenology, topography, parent material, and soil variables have been involved. Importance and capabilities of these variables for crop mapping have been discussed. However, all climate variables have not shown noticeable importance but some of the climate variables scored well in importance measures. In combination with other variables, it decreases OOB error rate (0.028) and vice versa, increases the potential of RF classifier. Variables associated with spectral responses (Landsat-8 derived) have shown greater importance in comparison to topographic (DEM derived) and climate variables. Two SAR data of different polarization modes have not shown their usefulness in model development.

The separability analysis using JM method has shown a better separation between almost all classes. Performance is scaled between excellent to less than poor performance for each class as well as overall. Each parameter is assigned with a class value corresponding to performance level. The score is calculated on the basis of class value to rank the algorithm. On the basis of class specific measures and as well as in multiclass measures, RF scored highest. Total time is also considered as one parameter for performance analysis. Overall RF performed best followed by gradient boosting. Other ensemble methods, ARCing, Bagging, and ISLE, are in decreasing order of performance. Non-ensemble method, decision tree performed better than ISLE.

References

An, Q., & Yang, B. (2007). A multicrop identification model based on stepwise removal learning-support vector machine using remote sensing images. *New Zealand Journal of Agricultural Research, 50*(5), 1013–1019.

Arenas-Toledo, J. M., & Epiphanio, J. C. N. (2011). Harmonic amplitude-terms mask to highlight agriculture in the savanna domain below the Brazilian Amazonian frontier. *International Journal of Remote Sensing, 32*(18), 5021–5034.

Breiman, L. (1996). Bagging predictors. *Machine Learning, 24*(2), 123–140.

Breiman, L. (1997). Arcing the edge. *Statistics, 4*, 1–14.

Breiman, L. (1998). Arcing classifier (with discussion and a rejoinder by the author). *The Annals of Statistics, 26*(3), 801–849.

Breiman, L. (2001). Randomforest. *Machine Learning, 45*(1), 5–32.

Breiman, L., Friedman, J. H., Stone, C. J., & Olshen, R. A. (1984). *Classification and regression trees*. Monterey, California, USA: Wadsworth and Brooks/Cole.

Brisco, B., Brown, R. J., Hirose, T., McNairn, H., & Staenz, K. (1998). Precision agriculture and the role of remote sensing: A review. *Canadian Journal of Remote Sensing, 24*(3), 315–327.

ChenChi, F., ChenCheng, R., & Son Nguyen, T. (2012). Investigating rice cropping practices and growing areas from MODIS data using empirical mode decomposition and support vector machines. *GIScience & Remote Sensing, 49*(1), 117–138.

Deschamps, B., McNairn, H., Shang, J., & Jiao, X. (2012). Towards operational radar-only crop type classification: Comparison of a traditional decision tree with a random forest classifier. *Canadian Journal of Remote Sensing, 38*(1), 60–68.

Dietterich, T. G. (2000). An experimental comparison of three methods for constructing ensembles of decision trees. *Machine Learning, 40*, 139–157.

Freund, Y., & Schapire, R. E. (1996). Experiments with a New Boosting Algorithm. In L. Saitta (ed.), *Proceedings of the Thirteenth International Conference on Machine Learning* (pp. 148–165). Bari, Italy: Morgan Kaufmann.

Freund, Y., & Schapire, R. E. (1997). A desicion-theoretic generalization of on-line learning and an application to boosting. *Journal of Computer and System Sciences, 55*, 119–139.

Friedman, J. H. (2001). Greedy function approximation: A gradient boosting machine. *The Annals of Statistics, 29*(5), 1189–1232.

Friedman, J. H., & Popescu, B. E. (2003). Importance sampled learning ensembles. *Computing, 94305*(2), 1–32.

Friedman, J. H., Hastie, T., & Tibshirani, R. (2000). Additive logistic regression: A statistical view of boosting (with discussion and a rejoinder by the authors). *The Annals of Statistics, 28*(2), 337–407.

Genuer, R., Poggi, J., & Tuleau-Malot, C. (2010). Variable selection using random forests. *Pattern Recognition Letters, 31*(14), 2225–2236.

Haboudane, D., Miller, J. R., Tremblay, N., Zarco-Tejada, P. J., & Dextraze, L. (2002). Integrated narrow-band vegetation indices for prediction of crop chlorophyll content for application to precision agriculture. *Remote Sensing of Environment, 81*(2–3), 416–426.

Harris, R. (2003). Remote sensing of agriculture change in Oman. *International Journal of Remote Sensing, 24*(23), 4835–4852.

Jaccard, P. (1912). The distribution of the flora in the alpine zone. *New Phytologist, 11*(2), 37–50.

Kumar, P., Gupta, D. K., Mishra, V. N., & Prasad, R. (2015). Comparison of support vector machine, artificial neural network, and spectral angle mapper algorithms for crop classification using LISS IV data. *International Journal of Remote Sensing, 36*(6), 1604–1617.

Kumar, P., Prasad, R., Choudhary, A., Mishra, V. N., Gupta, D. K., & Srivastava, P. K. (2016). A statistical significance of differences in classification accuracy of crop types using different classification algorithms. *Geocarto International, 6049*(May), 1–19.

Labatut, V., & Cherifi, H. (2011). Evaluation of performance measures for classifiers comparison. *Ubiquitous Computing and Communication Journal, 6*, 21–34.

Labatut, V., & Cherifi, H. (2012). Accuracy measures for the comparison of classifiers. In *The 5th International Conference on Information Technology* (p. 11). Amman, Jordan.

Liaw, A., & Wiener, M. (2002). *Classification and Regression by Random Forest, 2*(3), 18–22.

Mathur, A., & Foody, G. M. (2008). Crop classification by support vector machine with intelligently selected training data for an operational application. *International Journal of Remote Sensing, 29*(8), 2227–2240.

Mellor, A., Haywood, A., Stone, C., & Jones, S. (2013). The performance of random forests in an operational setting for large area Sclerophyll forest classification. *Remote Sensing, 5*(6), 2838–2856.

Metternicht, G. (2003). Vegetation indices derived from high-resolution airborne videography for precision crop management. *International Journal of Remote Sensing, 24*(14), 2855–2877.

Murthy, C. S., Raju, P. V., & Badrinath, K. V. S. (2003). Classification of wheat crop with multi-temporal images: Performance of maximum likelihood and artificial neural networks. *International Journal of Remote Sensing, 24*(23), 4871–4890.

Punera, K., & Ghosh, J. (2008). Consensus-based ensembles of soft clusterings. *Applied Artificial Intelligence, 22*(7–8), 780–810.

Rätsch, G., Onoda, T., & Müller, K. R. (2001). Soft margins for AdaBoost. *Machine Learning, 42*(3), 287–320.

Richards, J. A. (2013). Interpreting images. In *Remote sensing digital image analysis* (pp. 79–97). Berlin, Heidelberg: Springer Berlin Heidelberg.

Rilwani, M. L., & Ikhuoria, I. A. (2011). Prospects for geoinformatics-based precision farming in the Savanna River basin Nigeria. *International Journal of Remote Sensing, 32*(12), 3539–3549.

Rodriguez-Galiano, V. F., & Chica-Rivas, M. (2014). Evaluation of different machine learning methods for land cover mapping of a mediterranean area using multi-seasonal landsat images and digital terrain models. *International Journal of Digital Earth, 7*(6), 492–509.

Sarkar, A., Majumdar, A., Chatterjee, S., Chatterjee, D., Ray, S., & Kartikeyan, B. (2008). Study of the potential of alternative crops by integration of multisource data using a neuro-fuzzy technique. *International Journal of Remote Sensing, 29*(19), 5479–5493.

Seni, G., & Elder, J. F. (2010). *Ensemble methods in data mining: Improving accuracy through combining predictions*. Synthesis Lectures on Data Mining and Knowledge Discovery.

Sonobe, R., Tani, H., Wang, X., Kobayashi, N., & Shimamura, H. (2014). Parameter tuning in the support vector machine and random forest and their performances in cross- and same-year crop classification using TerraSAR-X. *International Journal of Remote Sensing, 35*(23), 7898–7909.

Sonobe, R., Tani, H., & Wang, X. (2016). An experimental comparison between KELM and CART for crop classification using Landsat-8 OLI data. *Geocarto International, 32*(2), 1–11.

Steinberg, D., & Colla, P. (1995). *CART: Tree-structured non-parametric data analysis*. San Diego, CA: Salford Systems

Tan, C. P., Ewe, H. T., & Chuah, H. T. (2011). Agricultural crop-type classification of multi-polarization SAR images using a hybrid entropy decomposition and support vector machine technique. *International Journal of Remote Sensing, 32*(22), 7057–7071.

Tumer, K., & Oza, N. C. (2003). Input decimated ensembles. *Pattern Analysis and Applications, 6*(1), 65–77.

Wang, X., Liang, T., Xie, H., Huang, X., & Lin, H. (2016). Climate-driven changes in grassland vegetation, snow cover, and lake water of the Qinghai Lake basin. *Journal of Applied Remote Sensing, 10*(3), 036017.

Weigend, A.S., Mangeas, M., & Srivastava, A. N. (1995). Nonlinear gated experts for time series: Discovering regimes and avoiding overfitting. *International Journal of Neural Systems*, 6(4), 373–399.

Chapter 6
Spatial Soil Moisture Prediction Model Over an Agricultural Land

Satellite remote sensing observations have the potential for efficient and reliable mapping of spatial soil moisture distributions. However, soil moisture retrievals from active/passive microwave remote sensing data are typically complex due to inherent difficulty in characterizing interactions among land surface parameters that contribute to the retrieval process. Therefore, adequate physical mathematical descriptions of microwave backscatter interaction with parameters such as land cover, vegetation density, and soil characteristics are not readily available. In such condition, non-parametric models could be used as a possible alternative for better understanding the impact of variables in the retrieval process and relating it in the absence of exact formulation. The following chapter will provide a conceptual framework for SMPM: A non-parametric approach over an agriculture land.

6.1 Introduction

Soil moisture is an important variable in land–atmosphere feedbacks at both weather and climate time scales. Global Climate Observing System (GCOS) has also endorsed soil moisture as one of the Essential Climate Variables (ECV) with others (GCOS-ECV 2010). Soil moisture is closely involved in the feedback between climate and vegetation, since both local climate and vegetation influence soil moisture through evapotranspiration, while soil moisture is a determinant of the type and condition of vegetation in a region. This means changes in soil moisture can accordingly have a substantial impact on agricultural productivity, forestry, and ecosystem health. (GCOS-195 2015).

Conventionally, in situ soil moisture measurements can be done through either direct or indirect method. These methods are cost prohibitive, time and resource consuming. Therefore research is oriented towards physical and empirical models. Physical models involve complicated scattering phenomena from probabilistic

P. K. Garg et al., *Digital Mapping of Soil Landscape Parameters*, Studies in Big Data 72,
https://doi.org/10.1007/978-981-15-3238-2_6

models of soil, vegetation, and terrain whereas empirical models capture relationships among measured variables to estimate geophysical characteristics (Ahmad et al. 2010). These models use precipitation, thermal index, crop water stress, theoretical functions of heat transfer and crop water stress for soil moisture study. With advancement in remote sensing techniques, a lot of research has been concentrated in this area based on the facts that reflected electromagnetic spectrum in the visible and infrared range is highly correlated to soil properties like soil colour, texture, surface roughness, silt, phosphorus and organic matter content, soil pH, available nitrogen, vegetation composition, as well as soil moisture (Gessler et al. 1995; Iverson et al. 1997; Morris and Boerner 1998; Hutchinson et al. 1999; Boerner et al. 2000; Hulley et al. 2010). Optical data derived indices such as Shortwave Angle Slope Index (Palacios-Orueta et al. 2005; Khanna et al. 2007), Soil Wetness Index (Giraldo et al. 2008; Mallick et al. 2009), soil evaporation transfer coefficient (Zhao et al. 2010) are also used for soil moisture profiling or measurement. They are often empirical and dependent on local meteorological conditions, such as wind speed, air temperature, and humidity (Mallick et al. 2009). The backscatter measured by radar and the emission measured by a radiometer are both very sensitive to the moisture content of bare-soil surfaces due to large contrast between dielectric properties of water and soil. In addition, microwave spectrum has soil penetration capability. In active microwave remote sensing, the radar backscattered signal from a bare soil depends on combination of factors, including radar properties (frequency, polarization), incident angle, and most importantly the composition and degree of heterogeneity of the surface (Ulaby et al. 1979; Dobson and Ulaby 1981; Du et al. 2000; Hsieh 2001; Verhoest et al. 2008; Wang and Qu 2009). Furthermore, if the soil is vegetated, additional information about biomass, canopy structure is needed (Ulaby et al. 1983; Wigneron et al. 2003; Loew 2008). Vegetation cover complicates the scattering and emission processes, by reducing the radiometric and radar soil-moisture sensitivities (Du et al. 2000). To reveal worthy information about soil moisture, the model must correct all influencing parameters. A lot of research has been concentrated in this area during the last two decades (Wigneron et al. 2003; Loew 2008).

From available research, it can certainly be stated that microwave range is most suitable for soil moisture studies if land is bare (Dobson and Ulaby 1981; Altese et al. 1996; Su et al. 1997; Shi et al. 1997; Li and Islam 1999; Du et al. 2000; Wigneron et al. 2003; Chai et al. 2009; Genis et al. 2013; Lakshmi 2013; Das and Paul 2015; MirMazloumi and Sahebi 2016) and visible and infrared range is most suitable for vegetation cover. Most of the works in this area are using optical images since 1972, year of first Landsat launch (if what is beneath the vegetation cover is not matter) (Purevdorj et al. 1998; Laamanen and Verlinden 2006; Buyantuyev et al. 2007; Muramatsu et al. 2007; Chen and Gillieson 2009; Gill et al. 2010, 2017; Townshend et al. 2012; Munyati and Sinthumule 2013; Desalegn et al. 2014; Li et al. 2015; Feng et al. 2016; Shukla et al. 2017) and for soil composition or soil texture analysis (if land is open or bare) (Gessler et al. 1995; Iverson et al. 1997; Morris and Boerner 1998; Hutchinson et al. 1999; Hulley et al. 2010). This study has combined these data set in such a way wherever they are most suited. In recent years, machine learning algorithms have received increasing attention for soil moisture mapping. Machine learning may be defined as the automated process to get the unknown relationship between input (predictor variable) and target parameters using a known example (training data), where the relationship may then be used on new data (testing data) for the purposes of prediction (Witten & Frank 2005). The targets can be nominal or numerical. If the targets are nominal, then the problem becomes a classification one while if the target is numerical, the problem is a regression one (Gorthi 2011). The learning task that involves solving such a regression problem is called supervised learning. Different machine learning classification/regression algorithms have been adopted for digital soil moisture mapping (DSMM) viz. CDT (Waldteufel et al. 2007; Myers et al. 2009), SVM (Yang and Huang 2002; Liu et al. 2008, 2010; Ahmad et al. 2010; Deng et al. 2011; Rodriguez-galiano et al. 2012; Zaman et al. 2012; Srivastava et al. 2013), RVM (Zaman et al. 2012), ANN (Dawson et al. 1997; Satalino et al. 2002; Jiang and Cotton 2004; Liu et al. 2008; Said et al. 2008; Chai et al. 2009; Baghdadi et al. 2012; Paloscia et al. 2013; Xie et al. 2014; Gao et al. 2014; Cui et al. 2016), Deep learning (Song et al. 2016; Zhang et al. 2017), KNN algorithm (Coopersmith et al. 2016) and different comparative studies are also presented (Lakhankar et al. 2009; Pasolli et al. 2012; Srivastava et al. 2013).

The surface soil moisture retrieval process is typically a challenging task, and it falls into the category of an ill-posed problem. The surface soil moisture under vegetation cover is three-layer problem. First is vegetation cover, second is soil composition and heterogeneity, and third is surface soil moisture. It can be stated that vegetation over surface and soil type have an influence on surface soil moisture. Therefore, the selection of important parameter representing vegetation cover and soil type is an important task. MDA and MDG score of RF algorithm has the capability to rank the potential variables (Breiman 2001). This study utilized MDA and MDG score to select effective indicators. Initially to incorporate dependency of crops and soil type on different variables namely, texture, phenological, parent material and soil, soil moisture, topographic, vegetation, and climate, 35 digital layers are prepared using different satellite data and climatic data (Chap. 3). The variables above the cut-off range (average value in this case) are selected as important parameters or representative of soil/crop class. To introduce large class variability of crop and soil, a larger area (2,421 km^2, constituting 11 districts, named as Site II) under an aridic to ustic climate zone is selected. Due to field observation limitation to conduct field experiments synchronous with satellite passes, a subset of this area (Site I) is selected for surface soil moisture modelling. Random sampling approach is adopted to collect soil samples at 0–5 cm depth of soil surface in six fields of Site I. Gravimetric method is used for in situ surface soil moisture measurements. Total 17 fields are selected in Site II for crop sampling. Extensive field survey about crop types in *rabi* season has been carried out to prepare training and testing data of crop cover (Shukla et al. 2017, 2018). The crop class samples have been also collected using random sampling approach. Pre-existing district soil resource maps (SRM) on 1:50,000 scale, provided by National Bureau of Soil Survey and Land Use Planning, India are used to obtain reference data about soil classes across the area. Stratified random sampling scheme is used to collect reference samples from pre-existing soil maps. Once the important variables are identified (or class representative of soil and crop), the different combination with backscattered data are used for soil moisture modelling. Three machine learning regression methods, RF, SVM and ANN, are used for model development and for comparative performance evaluation.

6.2 Modelling Framework

To fulfil the objective to develop a model for the quantitative measurement of surface soil moisture, the work in this chapter can be summarized as, Step 1, independent study on soil class and crop cover are carried out using RF classification algorithm, and importance of variables is calculated based on MDG and MDA score (Chaps. 4 and 5, respectively), and the cut-off is set to select most influential variables to represent crop and soil class variability; Step 2, the different combinations with backscattered data are used for soil moisture modelling using regression techniques RF, SVM, and ANN, model prediction in Site I is compared with field sample data; Step 3, model performance on the 9 km modelling grid is also evaluated with SMAP L4_SM data in the Site II. Modelling approach and the flow chart of SMPM are presented in Figs. 6.1 and 6.2, respectively.

6.2.1 Modelling Methods

Three machine learning regression methods RF, SVM, and ANN are used for model development and for comparative performance evaluation. A brief description of the principles of RF is provided in (Chap. 5, Sect. 5.2.1). Breiman et al. (1984), Breiman (2001) provide more details of RF. This section explains the theoretical mathematical backgrounds involved in developing SVM and ANN. A more detailed description of SVM and ANN can be found in Vapnik (2000) and (ASCE Task Committee 2000a, b), respectively.

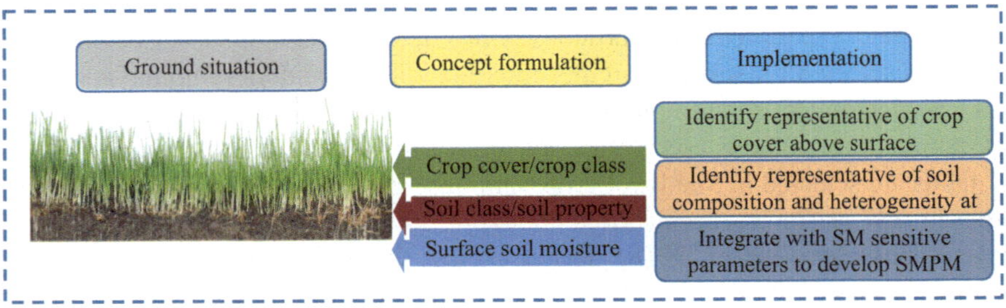

Fig. 6.1 SMPM modelling approach

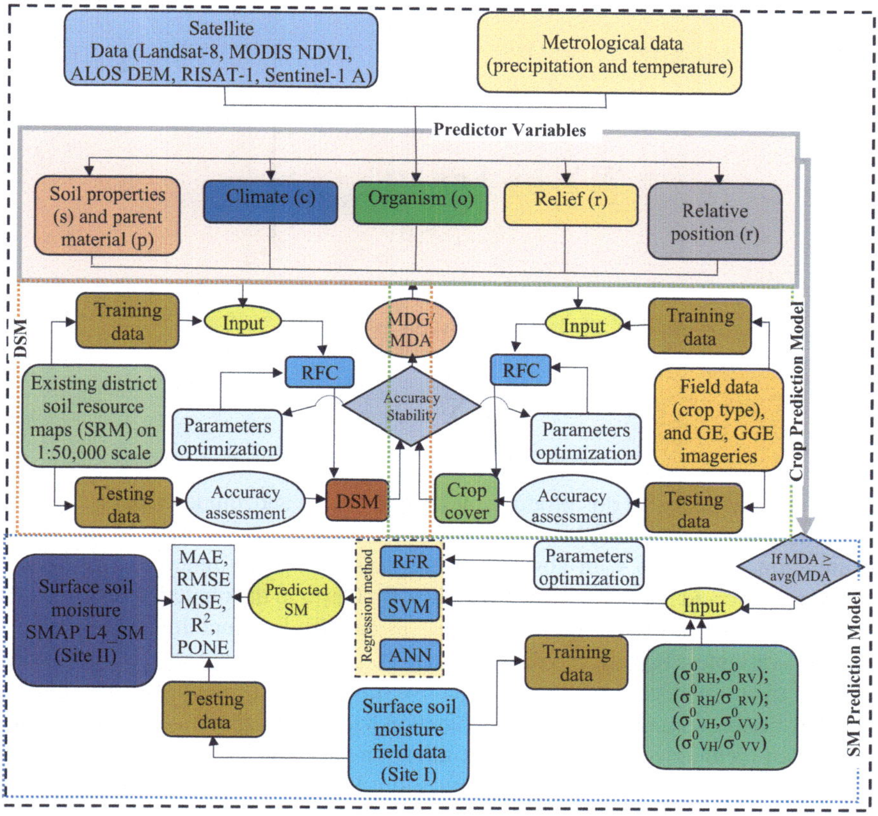

Fig. 6.2 Flow chart of SMPM

6.2.1.1 Support Vector Machine (SVM)

Vapnik and his co-workers developed SVMs for regression (Vapnik 2000). The functional dependency, $f(\mathbf{x})$, between independent variables, $x = \{x_1, x_2, \ldots, x_L\}$ where $x \in R^K$, and (dependent) variable, $y = \{y_1, y_2, \ldots, y_L\}$, where $y \in R$, is learned through the regularized functional:

$$\text{Minimize}\ \frac{1}{2}W^2 + C\sum_{i=1}^{L}(\xi_i + \xi_i^*) \tag{6.1}$$

Subject to

$$\left.\begin{array}{c} y_i - \sum_{j=1}^{K}\sum_{i=1}^{L} w_j x_{ji} - b \le \varepsilon + \xi_i; \\ \sum_{j=1}^{K}\sum_{i=1}^{L} w_j x_{ji} + b - y_i \le \varepsilon + \xi_i^*; \\ \xi_i, \xi_i^* \ge 0 \end{array}\right\} \tag{6.2}$$

where $i = 1, 2, \ldots, L$ is the sample size and $f(x) = w, x + b$, with $w \in x$ and $b \in R$, where, R is a real number. $f(x)$ can be written as follows:

$$f(x) = \sum_{i=1}^{K} w_j x_j + b \tag{6.3}$$

where w, x denotes the dot product of weight vector and input matrix, K is the dimension of w and x (number of support vectors) and b is the bias. In the above formulation, the errors with absolute values less than ε are ignored, making the solution sparse and hence 'ε-insensitive'. The first term in the objective function is a regularization term; the second term is the ε-insensitive loss function. The ξ_i and ξ_i^* are slack variables. For errors smaller than ε, ξ_i, and ξ_i^* are not required to be non-zero and the point does not enter the objective function. The quantity C controls the trade-off between minimizing the loss function and minimizing model complexity.

Another technique of solving the optimization problem, subject to constraints in the loss function, is using the dual formulation. In the dual formulation, Lagrange multipliers, α and α^* are introduced and the minimization equation is solved by differentiating with respect to the primary variables. It results in a maximization problem, i.e.

$$w(\alpha, \alpha^*) = \varepsilon \sum_{i=1}^{L}(\alpha_i + \alpha_i^*) + \sum_{i=1}^{L} y_i(\alpha_i - \alpha_i^*) - \frac{1}{2}\sum_{i,j=1}^{L}(\alpha_i - \alpha_i^*)\left(\alpha_j - \alpha_j^*\right)k(x_i, x_j) \tag{6.4}$$

Subject to

$$\sum_{i=1}^{L}(\alpha_i^* - \alpha_i) = 0; 0 \le \alpha_i, \alpha_i^* \le C, \tag{6.5}$$

and the approximating function $f(x)$

$$f(x) = \sum_{i=1}^{N}(\alpha_i^* - \alpha_i)k(x, x_i) + b \tag{6.6}$$

where α^* and α are Lagrange multipliers and $k(x, x_i)$ is the radial basis kernel function, which has the form $k(x_i, x_j) = \exp\left(-\gamma |x_i - x_j|^2\right)$ wherein γ is the kernel parameter, which is selected on the basis of a trial-and-error procedure. The x_i's are 'support vectors' (SVs) and N is the number of SVs corresponding to values of the independent variables that are at least ε away from actual observations. For $|f(x_i) - y_j| \geq \varepsilon$, the Lagrange multipliers are non-zero, and for the points inside the ε-tube, the parameters α_i^*, α_i vanish.

6.2.1.2 Artificial Neural Network (ANN)

The ANN model developed corresponds to feedforward–backpropagation type of models. The feedforward–backpropagation is adapted due to its applicability in a variety of different problems (ASCE Task Committee 2000a, b; Ahmad et al. 2010). A two-layer feedforward network with sigmoid hidden neurons and linear output neurons can fit multi-dimensional mapping problems arbitrarily well, given consistent data and enough neurons in its hidden layer. Feedforward networks often have one or more hidden layers of sigmoid neurons followed by an output layer of linear neurons. Multiple layers of neurons with nonlinear transfer functions allow the network to learn nonlinear relationships between input and output vectors. The network is trained with Levenberg–Marquardt backpropagation algorithm. This algorithm typically requires more memory but less time. Training automatically stops when generalization stops improving, as indicated by an increase in the mean square error of the validation samples.

6.2.2 Model Development and Implementation

All modelling is performed in Matlab (MATLAB®). The randomForest-matlab (@misc) package is used to build the RF models (classification and regression). LIBSVM library is used for SVM (Chang and Lin 2011). Neural Network Toolbox™ provides simple MATLAB® commands for creating ANN architecture, therefore used to create ANN models in this study. Detailed information of methods is presented in the previous section.

The RF code is based on R source by Liaw and Wiener (2002) of RF algorithm, which is originally developed by Leo Breiman and Adele Cutler in FORTRAN (Breiman 2001; Liaw and Wiener 2002). To find the ideal value of parameters *ntree* and *mtry*, the error rate is calculated in OOB and test data set for different *mtry,* i.e. D_{min}, D_{max}, D/2, $\sqrt{D}$ with 1000 trees, where D = number of predictor variables. The value of *mtry* at which error is minimum and stable is considered as the optimum value of *mtry*, and corresponding value of *ntree* is considered as optimum number of tree in RF model development. SVM models are also tuned using optimum value of epsilon and cost function. The tuning method is known as grid search in LIBSVM, which runs various models with all the possible values of epsilon and cost function in the specified range and gives the model which has the lowest error.

If the targets are nominal, then the problem becomes a classification one, while if the target is numerical, the problem is a regression one. Therefore in identification of important variables, RF model is used as a classifier for soil and crop class in agriculture land. Once the important variables are identified, the different combination with backscattered data is used for soil moisture modelling. RF model prediction using σ^0_{RH} and σ^0_{RV} with other variables denoted by RF(σ^0_{RH}, σ^0_{RV}), and using $\sigma^0_{RH}/\sigma^0_{RV}$ with other variables denoted by RF($\sigma^0_{RH}/\sigma^0_{RV}$) throughout this chapter. Likewise, RF model prediction using σ^0_{VH} and σ^0_{VV} with other variables denoted by RF(σ^0_{VH}, σ^0_{VV}) and using $\sigma^0_{VH}/\sigma^0_{VV}$ with other variables denoted by RF($\sigma^0_{VH}/\sigma^0_{VV}$) throughout this chapter. Similar conventions have been followed for other models. Therefore, four models for each methods are RF(σ^0_{RH}, σ^0_{RV}), RF($\sigma^0_{RH}/\sigma^0_{RV}$), RF($\sigma^0_{VH}$, σ^0_{VV}), RF($\sigma^0_{VH}/\sigma^0_{VV}$) for RF method; SVM($\sigma^0_{RH}$, σ^0_{RV}), SVM($\sigma^0_{RH}/\sigma^0_{RV}$), SVM($\sigma^0_{VH}$, σ^0_{VV}), SVM($\sigma^0_{VH}/\sigma^0_{VV}$) for SVM method; ANN($\sigma^0_{RH}$, σ^0_{RV}), ANN($\sigma^0_{RH}/\sigma^0_{RV}$), ANN($\sigma^0_{VH}$, σ^0_{VV}), ANN($\sigma^0_{VH}/\sigma^0_{VV}$) for ANN method. These all models are tested for surface soil moisture content prediction in Site I. Model performance on the 9 km modelling grid is also evaluated. However, based on their satisfactory performance in ground testing samples on Site I, only four models are chosen to evaluate for L4_SM data in Site II, i.e. RF(σ^0_{RH}, σ^0_{RV}), RF($\sigma^0_{RH}/\sigma^0_{RV}$), RF($\sigma^0_{VH}$, σ^0_{VV}), and SVM(σ^0_{RH}, σ^0_{RV}).

6.2.3 Identification of Effective Indicators for Soil and Cover

This study considered surface soil moisture in agriculture land is three-layer problem. First is vegetation cover (crop), second is soil composition and heterogeneity, and third is surface soil moisture. It can be stated that vegetation over surface and soil type has the influence on surface soil moisture. Therefore to identify variables to represent vegetation cover and soil surface (composition and heterogeneity), independent study is performed using RF classification. Detailed work of soil class and crop cover mapping is discussed in previous chapters (Chaps. 4 and 5, respectively). To introduce large class variability of crop and soil, a larger area (2,421 km^2, constituting 11 districts, named as Site II) under an aridic to ustic climate zone is selected. RF algorithm has the capability to rank the variables (Breiman 2001) based on MDA and MDG score. Therefore, MDA and MDG score are utilized to select important variables in both cases and considered these as representative of vegetation and soil. Such estimation is of importance for variable selection in the classification of complex areas where it is mandatory to use large multisource data sets with a large number of variables. Before variable selection, RF models are tuned by using the ideal value of their two parameters based on error stability in OOB data.

In previous studies on sol and crop mapping, RF models of soil and crop study are trained using training data to predict classes. Then models are evaluated using testing data set. Hence, the model performance depends on variable selection. Therefore, good performance of classifier for these studies indicated that the selected variables are a good representative of the class. However, it is necessary to identify variables, those who have significant contribution in model development. Therefore, MDA (Fig. 6.3a) and MDG (Fig. 6.3b) scores are calculated. Gini importance is not fair in favour of predictor variables with many categories, while the RF permutation importance is a more reliable indicator (Strobl et al. 2007).

Average value of score is considered as cut-off value for selection of variables of major importance. The variables above the cut-off value are presented with light green and reddish brown colour within their textured bar for crop and soil class mapping, respectively, in Fig. 6.3.

With the high MDA and MDG score in both classes (soil and crop), Vegetation related variables NDVI, texture measure in NDVI (Variance and Diversity) and phenological temporal variance in NDVI (NDVI variance) have shown their importance in digital mapping of soil as well as in crop cover. A large portion of the study area is

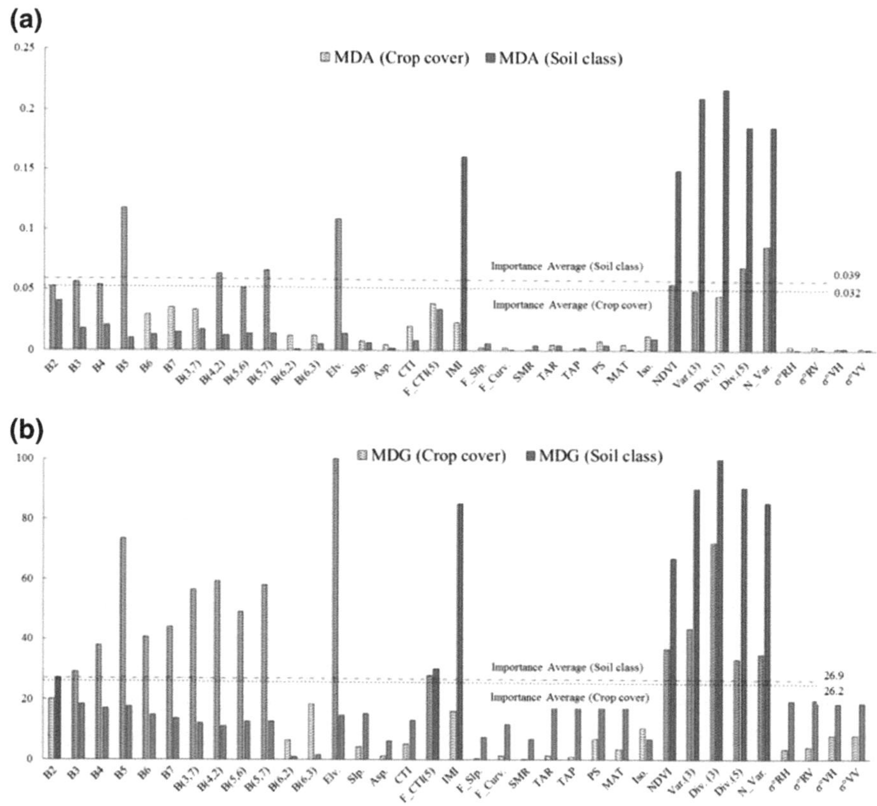

Fig. 6.3 **a** MDA score and **b** MDG score for crop and soil class study. Average value of score is considered as cut-off value for selection of variables of major importance. The variables above the cut-off value are presented with light green and reddish brown colour within their textured bar for crop and soil class mapping, respectively

agricultural land (93%). Many common agricultural practices, especially ploughing, disc-tillage, and vegetation burning, accelerate the decomposition and humification of soil organic matter and leave the soil susceptible to wind and water erosion. Thus have an influence on soil formation. Actually, the soil is the result of the combined activity of environment and terrain features. Therefore, it was expected to get a good score for vegetation related variables in both classes.

Landsat bands and their ratio also scored well in crop mapping. Landsat-8 band 5 (near-infrared band) is the most important predictor variable based on MDA and MDG scores in crop mapping. The RGB bands also scored higher than cut-off value. Out of six normalized difference band ratios variables, three variables: B5 – B7/B5 + B7, B4 – B2/B4 + B2, and B5 – B6/B5 + B6 have also scored higher than the cut-off value in both scores. It indicated that the distinct vegetation communities and landforms can easily be identified in different spectral bands as well as in different normalized difference band ratio. However, in soil mapping, they have not shown significant contribution.

In both scores, DEM derived topographical variables, slope, aspect, and their spatial context appeared to not have a significant impact on soil and crop cover mapping in the relatively flat surface. However, elevation is the second most important predictor variable in crop cover mapping. The study area is relatively flat (relative to nearest pixel) but the elevation values range from 90 to 160 m. Therefore it indicates that elevation has an impact on land features discrimination, i.e. crop and non-crop area in crop study.

The topographic variable associated with soil attributes, IMI, scored well in digital soil mapping. However, CTI has not shown as much importance as expected for digital soil mapping. The study area is relatively flat, and in flat area, accumulation numbers will be very large; therefore, CTI has not shown its relevance in relatively flat area. However, textured CTI scored well in MDG score. Although CTI has not scored well, but its texture measure scored significant value in the both scores, even its value is above the cut-off value in MDG score of soil and crop classes. It indicated the importance of spatial context in soil attributes. CTI and IMI have not shown their importance in crop mapping in this study area.

Although, in the local climate condition, soils and crops are different according to different climate regimes such as temperature and precipitation (McBratney et al. 2003; Ramírez-villegas and Bueno-cabrera 2009; Xin et al. 2015). But in this study, climate variables have not shown their importance as much as expected for class prediction in any of the important measurements in soil and crop mapping.

Similarly, the backscattering responses in two polarization modes have also scored lower values in MDA and MDG score in crop mapping. The backscatter response is a function of the crop roughness, moisture, and geometry and different crops have different backscattering properties based on their canopy structure and moisture content (Cable et al. 2014; Mishra et al. 2014; Liu et al. 2016a, b). This study has used C-band, which is most suitable for crop study in the agriculture land. However, the scores are low. In this study, both optical and SAR are used, and vegetation related variables derived from Landsat-8 data has already indicated the major impact on crop and soil mapping as discussed above. This should be the reason for low value in the importance measurement. Therefore it is assumed that backscattering response is dominated by optical data derived vegetation indices in crop mapping. It also indicates that optical data derived vegetation indices and their texture measures are the best representative of crop cover (or vegetation cover). However, in soil mapping, variables related to backscattering response have scored lower values in MDA but the comparatively significant value in MDG score, however, value is below the average value. Therefore it is concluded that backscattering responses do not have a significant impact on soil and crop mapping in this study.

It can be observed from Fig. 6.3 that most of the variables which scored higher than the cut-off value in MDA score, have also scored higher value in MDG score. However, in MDG score, some other variables have also shown their importance. Many researchers have shown that MDA score is a more reliable indicator than Gini importance (Strobl et al. 2007; Oppel et al. 2009; Altmann et al. 2010; Louppe et al. 2013). Therefore the variables that scored higher than the cut-off value in MDA score are only considered, and assumed that these variables are representative of their class.

Based on above, fourteen variables; Landsat-8 bands: B2, B3, B4, B5, band ratio: $B4-B2/B4+B2$, $B5-B6/B5+B6$, $B5-B7/B5+B7$, elevation, IMI, vegetation and their texture related variables: NDVI, Variance (3×3), Variance (5×5), Divergence (5×5), and phenological variable: NDVI variance is selected as a representative of crop cover and soil class in the agriculture land to develop soil moisture modelling.

6.2.4 *Model Evaluation*

6.2.4.1 Goodness-of-Fit Tests

To test the degree of association between the actual and predicted sample data, goodness-of-fit evaluation measures are used. This study adopted popular performance measures, i.e. mean absolute error (MAE), root mean squared error (RMSE), mean squared error (MSE), and mean absolute percentage error (MAPE), which are used to measure the magnitude of error (Patel et al. 2006; Ahmad et al. 2010; Zaman et al. 2012). RMSE is better for showing bigger deviations, MAE and MSE have easiest interpretation and MAPE measures the size of the error in percentage terms. MAPE is calculated as the average of the unsigned percentage error. These statistics are calculated as

$$\mathrm{MAE} = \frac{\sum_{i=0}^{N} |y_i - f(x)_i|}{N} \tag{6.7}$$

$$\mathrm{RMSE} = \sqrt{\frac{\sum_{i=0}^{N} (y_i - f(x)_i)^2}{N}} \tag{6.8}$$

$$\mathrm{MSE} = \frac{\sum_{i=0}^{N} (y_i - f(x)_i)^2}{N} \tag{6.9}$$

$$\text{MAPE} = \frac{100}{N}\left(\sum_{i=0}^{N}\left(\frac{|y_i - f(x)_i|}{y_i}\right)\right) \quad (6.10)$$

where N is the number of test samples. In this formulation, y and $f(x)$ are the actual and predicted values, respectively. MAE, RMSE, and MSE can range from 0 to ∞. For prediction which is too low, the MAPE cannot exceed 100%. Lower values of these measures are better. The square of the correlation (R^2) is also calculated. R^2 is a statistical measure of how close the data are to the fitted regression line. It is also known as the coefficient of determination, or the coefficient of multiple determination for multiple regression. It is always between 0 and 100%. The higher the R^2, the better the model fits your data.

6.2.4.2 Probability of Non-exceedance (PONE)

It shows the expected chance (x-axis, in percent) that the error in model prediction will not exceed (i.e. not be higher than) the error shown on the y-axis in percent. When the percent errors are arranged in ascending order, then the Probability of non-exceedance (PONE) can be estimated by

$$\text{PONE} = \frac{R}{N+1} \quad (6.11)$$

$$\%\,\text{error} = 100 \times \left(\frac{|y_i - f(x)_i|}{y_i}\right) \quad (6.12)$$

where N is the number of test samples and R is the rank number of the observation. The plot between percent error and PONE is used to show the probability (x-axis, in percent) that the error in model prediction will not exceed the error shown on the y-axis in percent.

6.3 Result and Discussion

6.3.1 Models Performance on Field Samples on Site I

The scatter plot between model predicted and actual surface soil moisture for all models are shown in Figs. 6.4 and 6.5. Performance measures are listed in Table 6.1. RMSE and R^2 values range from 0.93–0.87, 0.59–0.81 for RF models; 0.91–0.82, 0.70–0.99 for SVM models, and 0.86–0.51, 0.81–1.59 for ANN models, respectively. Results clearly indicate the satisfactory performance of RF and SVM for surface soil moisture retrieval. RF(σ^0_{RH}, σ^0_{RV}) model performed best (MSE = 0.003, RMSE

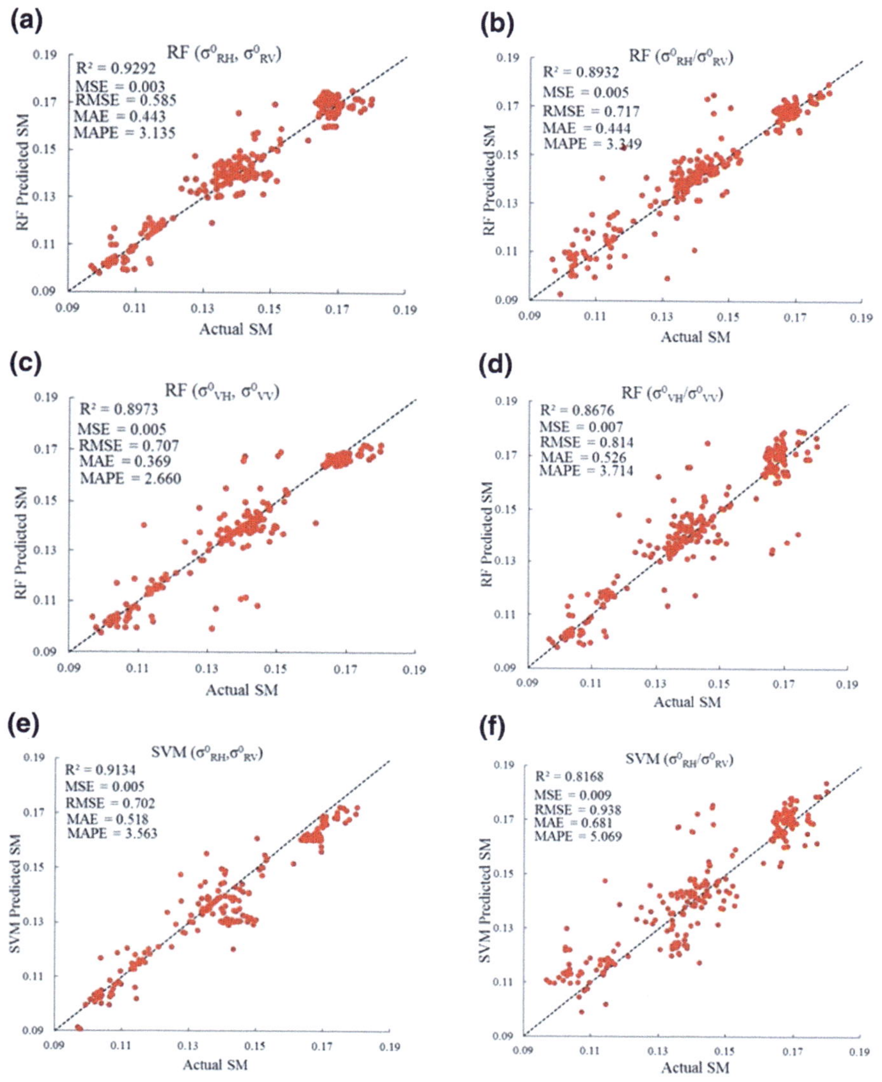

Fig. 6.4 Correlation between predicted and actual surface soil moisture content for model **a** RF with $\sigma^0{}_{RH}$, $\sigma^0{}_{RV}$; **b** RF with $\sigma^0{}_{RH}/\sigma^0{}_{RV}$; **c** RF with $\sigma^0{}_{VH}$, $\sigma^0{}_{VV}$; **d** RF with $\sigma^0{}_{VH}/\sigma^0{}_{VV}$; **e** SVM with $\sigma^0{}_{RH}$, $\sigma^0{}_{RV}$; **f** SVM with $\sigma^0{}_{RH}/\sigma^0{}_{RV}$. Dotted line is the 45° bisector line

= 0.59, MAE = 0.44, MAPE = 3.14, and R^2 = 0.93) in estimating the soil moisture content in the region. Other RF models; RF($\sigma^0{}_{RH}/\sigma^0{}_{RV}$), RF($\sigma^0{}_{VH}$, $\sigma^0{}_{VV}$), and RF($\sigma^0{}_{VH}/\sigma^0{}_{VV}$) are also performed satisfactorily (Table 6.1).

For high moisture content (0.16–0.19 cm^3), most of the sample points are concentrated around the 45° line (dotted line) in RF scatter plots, indicating that the model is trained well for high moisture content, i.e. more accurate prediction for

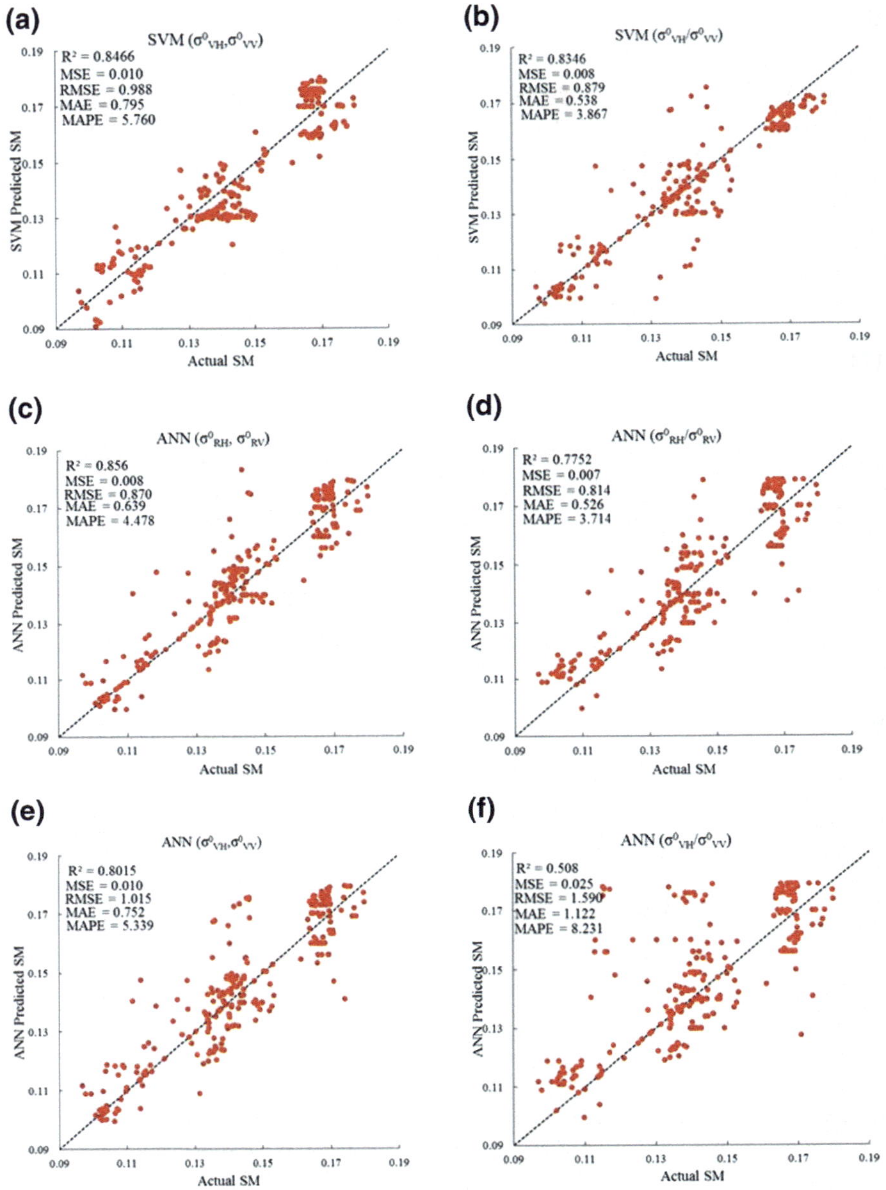

Fig. 6.5 Correlation between predicted and actual surface soil moisture content for model **a** SVM with σ^0_{VH}, σ^0_{VV}; **b** SVM with $\sigma^0_{VH}/\sigma^0_{VV}$; **c** ANN with σ^0_{RH}, σ^0_{RV}; **d** ANN with $\sigma^0_{RH}/\sigma^0_{RV}$; **e** ANN with σ^0_{VH}, σ^0_{VV}; **f** ANN with $\sigma^0_{VH}/\sigma^0_{VV}$. The dotted line is the 45° bisector line

Table 6.1 Performance measures of RF, SVM, and ANN models on testing samples on Site I. The MSE, RMSE, and MAE values are in percentage (%)

Site I var.	RF					SVM					ANN				
com	MSE	RMSE	MAE	MAPE	R^2	MSE	RMSE	MAE	MAPE	R^2	MSE	RMSE	MAE	MAPE	R^2
σ^0_{RH}, σ^0_{RV}	0.003	0.585	0.443	3.135	0.929	0.005	0.702	0.518	3.563	0.913	0.008	0.870	0.639	4.478	0.856
$\sigma^0_{RH}/\sigma^0_{RV}$	0.005	0.717	0.444	3.349	0.893	0.009	0.938	0.681	5.069	0.817	0.007	0.814	0.526	3.714	0.775
σ^0_{VH}, σ^0_{VV}	0.005	0.707	0.369	2.660	0.897	0.010	0.988	0.795	5.760	0.845	0.010	1.015	0.752	5.339	0.802
$\sigma^0_{VH}/\sigma^0_{VV}$	0.007	0.814	0.526	3.714	0.868	0.008	0.879	0.538	3.867	0.835	0.025	1.590	1.122	8.231	0.508

high moisture condition. In RF models, sample points corresponding to moderate (0.12–0.16 cm^3) and low (0.09–0.12 cm^3) moisture content have mixed behaviour; most of the points are concentrated around 45° line, but some sample points are scattered above and below the 45° line (Fig. 6.4a–d). A point lying far above the 45° line signifies higher predictions whereas a point far below the line signifies lower predictions. 40% testing samples are overestimated and 48% are underestimated in RF($\sigma^0{}_{RH}$, $\sigma^0{}_{RV}$) model.

For high moisture content, SVM($\sigma^0{}_{RH}$, $\sigma^0{}_{RV}$) model predicted lower than actual SM, for all sample points. However, for low to moderate soil moisture condition, points are scattered above and below the 45° line, similar to others SVM models. SVM($\sigma^0{}_{RH}$, $\sigma^0{}_{RV}$) model performed satisfactorily with MSE = 0.005, RMSE = 0.70, MAE = 0.52, MAPE = 3.56, and R^2 = 0.91. However, all others SVM models performance is not much impressive.

The noticeable results for ANN models is only for ANN($\sigma^0{}_{RH}$, $\sigma^0{}_{RV}$), which scored MSE = 0.008, RMSE = 0.87, MAE = 0.64, MAPE = 4.48, and R^2 = 0.86. All other ANN models show poor estimation of soil moisture content (Fig. 6.5c–f). Lowest value is for ANN($\sigma^0{}_{VH}/\sigma^0{}_{VV}$), i.e. MSE = 0.025, RMSE = 1.59, MAE = 1.12, MAPE = 8.23, and R^2 = 0.51.

The probability of Non-Exceedance (PONE) plot provides more detailed information about the reliability of model prediction (Fig. 6.6a–l). The horizontal light dotted line shows an error value of 5%, and the dark dotted horizontal line shows an error value of 10%. As one moves along each curve from left to right, the probability increases and the error also increases, indicating that it is more likely not to exceed higher error values. PONE plot of RF model for both combinations of $\sigma^0{}_{RH}$, $\sigma^0{}_{RV,}$ i.e. $\sigma^0{}_{RH}$, $\sigma^0{}_{RV;}$ $\sigma^0{}_{RH}/\sigma^0{}_{RV}$, indicates that there is 80% chance that the error will not be higher than 5%; and expects more than 90% chance for the error not to be higher than 10%. RF ($\sigma^0{}_{VH}/\sigma^0{}_{VV}$) model prediction also shows the nearly similar probability of non-exceedance for 5 and 10% error, however, RF($\sigma^0{}_{VH}$, $\sigma^0{}_{VV}$) model prediction exhibits more than 85% chance for the error not to be higher than 5%. The PONE plots of SVM models show that for model SVM($\sigma^0{}_{RH}$, $\sigma^0{}_{RV}$) and SVM($\sigma^0{}_{VH}$, $\sigma^0{}_{VV}$), approximately 75% of the estimates have an error of 5% or less which increases to 10% or less for more than 90% of the estimates. Although, SVM($\sigma^0{}_{RH}/\sigma^0{}_{RV}$) and SVM($\sigma^0{}_{VH}/\sigma^0{}_{VV}$) model show a high probability of non-exceedance ($\geq$90%) for 10% error, but for the error of 5%, models show less probability ($\leq$65%) that the error will not exceed than 5%.

PONE plots for ANN models also show a high probability, similar to RF and SVM, for the error of 10% but show very less chance for the error not to be higher than 5%. For ANN($\sigma^0{}_{VH}/\sigma^0{}_{VV}$) and ANN($\sigma^0{}_{VH}$, $\sigma^0{}_{VV}$), the probability is less than 40% that the error will not exceed 5%. Therefore, PONE plots indicate that for higher error value (10% or more), all models show approximate similar behaviour, but to get the high probability for the error not to exceed 5%, models selection, as well as parameters selection is an important task. Overall, all RF models show a higher probability of non-exceedance for less prediction error in comparison to SVM and ANN models. Based on the past research addressing the difficulties associated with measuring soil moisture in a comprehensive manner (Ulaby et al. 1979; Wagner

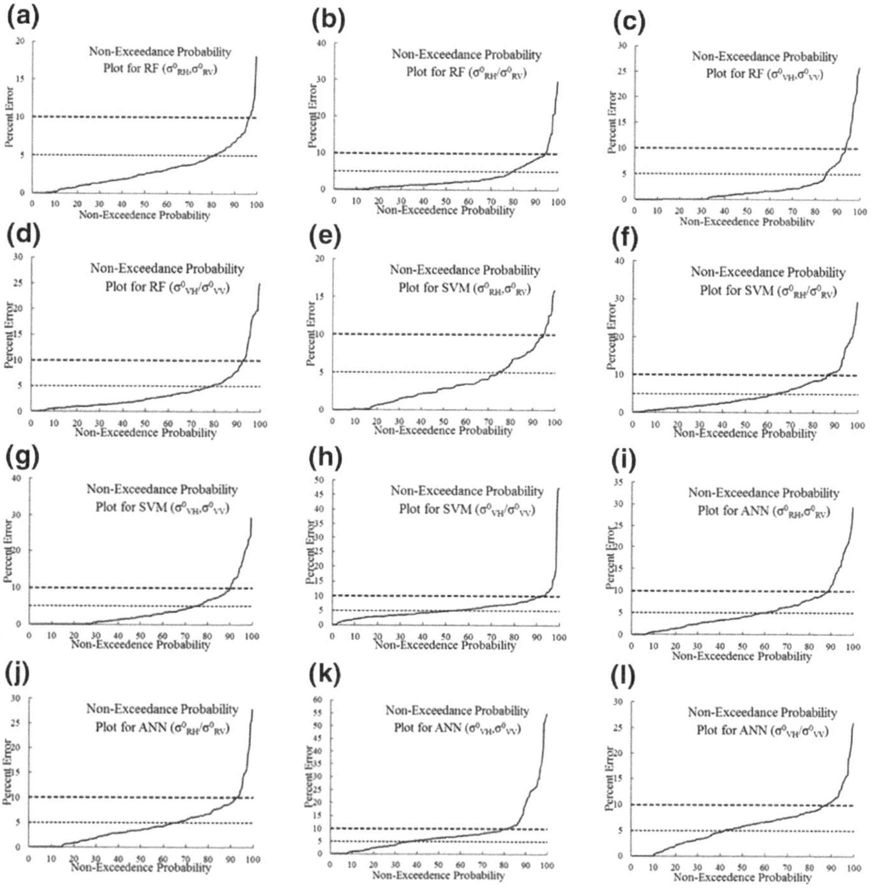

Fig. 6.6 Probability of non-exceedance for model **a** RF with σ^0_{RH}, σ^0_{RV}; **b** RF with $\sigma^0_{RH}/\sigma^0_{RV}$; **c** RF with σ^0_{VH}, σ^0_{VV}; **d** RF with $\sigma^0_{VH}/\sigma^0_{VV}$; **e** SVM with σ^0_{RH}, σ^0_{RV}; **f** SVM with $\sigma^0_{RH}/\sigma^0_{RV}$; **g** SVM with σ^0_{VH}, σ^0_{VV}; **h** SVM with $\sigma^0_{VH}/\sigma^0_{VV}$; **i** ANN with σ^0_{RH}, σ^0_{RV}; **j** ANN with σ^0_{RH}, σ^0_{RV}; **k** ANN with σ^0_{VH}, σ^0_{VV}; **l** ANN with $\sigma^0_{VH}/\sigma^0_{VV}$. The horizontal light dotted line shows an error value of 5% and the dark dotted horizontal line shows an error value of 10%

and Scipal 2000; Álvarez-Mozos et al. 2007; Paloscia et al. 2008; Said et al. 2008; Lakhankar et al. 2009; Gao et al. 2014), high probability of non-exceedance for low error value shows the robustness of the RF models in capturing the variability in soil moisture for the region.

Overall, the models seem to perform comparatively better at higher soil moisture values but do not show reasonable estimates at moderate and low soil moisture values. Most of the scattered samples are of moderate moisture value, therefore have major impact on model performance. One reason for unpredictability of moderate soil moisture value may be due to the mature crop cover at that region in the month, which dominates the backscatter signal. Major crop in this region is *mustard* and which is in maturity stage in March month. All other variables are the same during models' development, therefore it is expected that performance is majorly governed by backscattered measures. Although, this study attempts to develop spatial soil moisture prediction model over an agricultural land and results indicated overall satisfactory performance of RF and SVM($\sigma^0{}_{RH}$, $\sigma^0{}_{RV}$) for surface soil moisture retrieval. Additionally, within each machine learning method (RF, SVM, and ANN), model trained using $\sigma^0{}_{RH}$, $\sigma^0{}_{RV}$ scored higher value in comparison to others. PONE plot also showed a high probability of non-exceedance for low error value in this combination. Therefore, higher value of performance measures for models using σ0RH, σ0RV with other variables emphasized importance of circular polarized data in soil moisture prediction. Furthermore, models trained using backscattered bands scored higher value in comparison to their ratios. This result indicated that backscattered bands are more suitable than their ratios.

6.3.2 Model Performance on the 9 km Modelling Grid in Site II

Model performance on the 9 km modelling grid is also evaluated. Correlation between predicted surface soil moisture and SMAP L4_SM product is shown in Fig. 6.7a–d, for models RF($\sigma^0{}_{RH}$, $\sigma^0{}_{RV}$), RF($\sigma^0{}_{RH}/\sigma^0{}_{RV}$), RF($\sigma^0{}_{VH}$, $\sigma^0{}_{VV}$), and SVM($\sigma^0{}_{RH}$, $\sigma^0{}_{RV}$). Results at 9 km grid are also presented in Fig. 6.8. Only these four models are chosen to evaluate for SMAP L4_SM data, based on their satisfactory performance in ground testing samples on Site I. However, the models do not show similar performance as for Site I. But the performance measure values are up to satisfactory level for RF($\sigma^0{}_{RH}$, $\sigma^0{}_{RV}$) and RF($\sigma^0{}_{RH}/\sigma^0{}_{RV}$). RF($\sigma^0{}_{RH}$, $\sigma^0{}_{RV}$) shows correlation with MSE = 0.005, RMSE = 0.66, MAE = 0.53, MAPE = 4.20, and R^2 = 0.87 and RF($\sigma^0{}_{RH}/\sigma^0{}_{RV}$) with MSE = 0.006, RMSE = 0.76, MAE = 0.61, MAPE = 4.87, and R^2 = 0.87. R^2 value is only dropped by 6.5% for RF($\sigma^0{}_{RH}$, $\sigma^0{}_{RV}$), and by 2.2% for RF($\sigma^0{}_{RH}/\sigma^0{}_{RV}$) in comparison to the correlation between model predicted and ground samples. RF($\sigma^0{}_{VH}$, $\sigma^0{}_{VV}$) also shows good performance measures value, i.e. MSE = 0.006, RMSE = 0.74, MAE = 0.56, MAPE = 4.42, and R^2 = 0.85. Although SVM($\sigma^0{}_{RH}$, $\sigma^0{}_{RV}$) model has shown good correlation (R^2 = 0.91) between predicted and ground testing data, but predicted value does show a satisfactory correlation with L4_SM data at 9 km grid. The results indicated good prediction stability of RF method in comparison to SVM and good capability of capturing the variability in soil moisture. Models with circular polarized data show better capability in surface soil moisture prediction in comparison to linear, it also supports the previous findings.

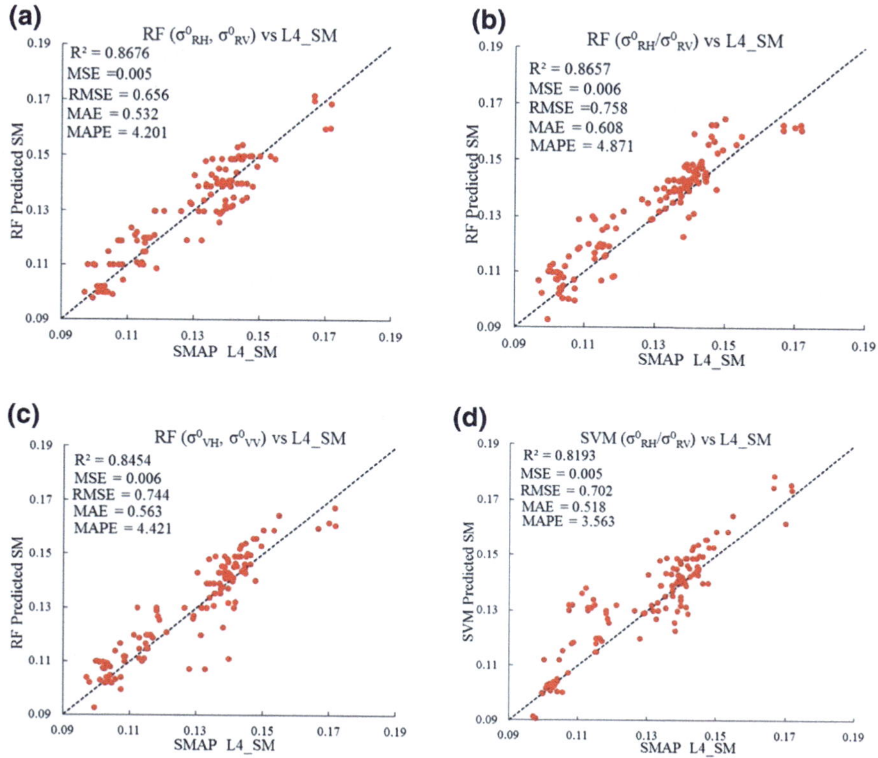

Fig. 6.7 Correlation between predicted and SMAP L4_SM, for model **a** RF with σ^0_{RH}, σ^0_{RV}; **b** RF with $\sigma^0_{RH}/\sigma^0_{RV}$; **c** RF with σ^0_{VH}, σ^0_{VV}; and **d** SVM with σ^0_{RH}, σ^0_{RV}. Dotted line is the 45° bisector line

6.4 Summary

This study provides insight into the use of prominent regression techniques, RF, SVM, and ANN as a spatial surface soil moisture model over an agricultural land. Most of the study area is agricultural land (93%), therefore, vegetation related variables NDVI, texture measure in NDVI (Variance and Diversity), and phenological temporal variance in NDVI (NDVI variance) are dominant variables in digital mapping of soil as well as in crop cover. Landsat-8 bands: B2, B3, B4, B5, band ratio: B4 – B2/B4 + B2, B5 – B6/B5 + B6, B5 – B7/B5 + B7, elevation, IMI also use full variables in soil and crop mapping. Based on an independent study on soil type and crop cover, fourteen variables were selected as representative of crop and soil class in the agricultural land. These variables were used with different backscattered responses to develop twelve digital surface soil moisture models. Goodness-of-fit tests indicate the degree of association between the actual and model's predicted surface soil moisture value for each developed model. Results clearly indicate the satisfactory performance of RF models, and SVM(σ^0_{RH}, σ^0_{RV}) for surface soil moisture retrieval. RF(σ^0_{RH},

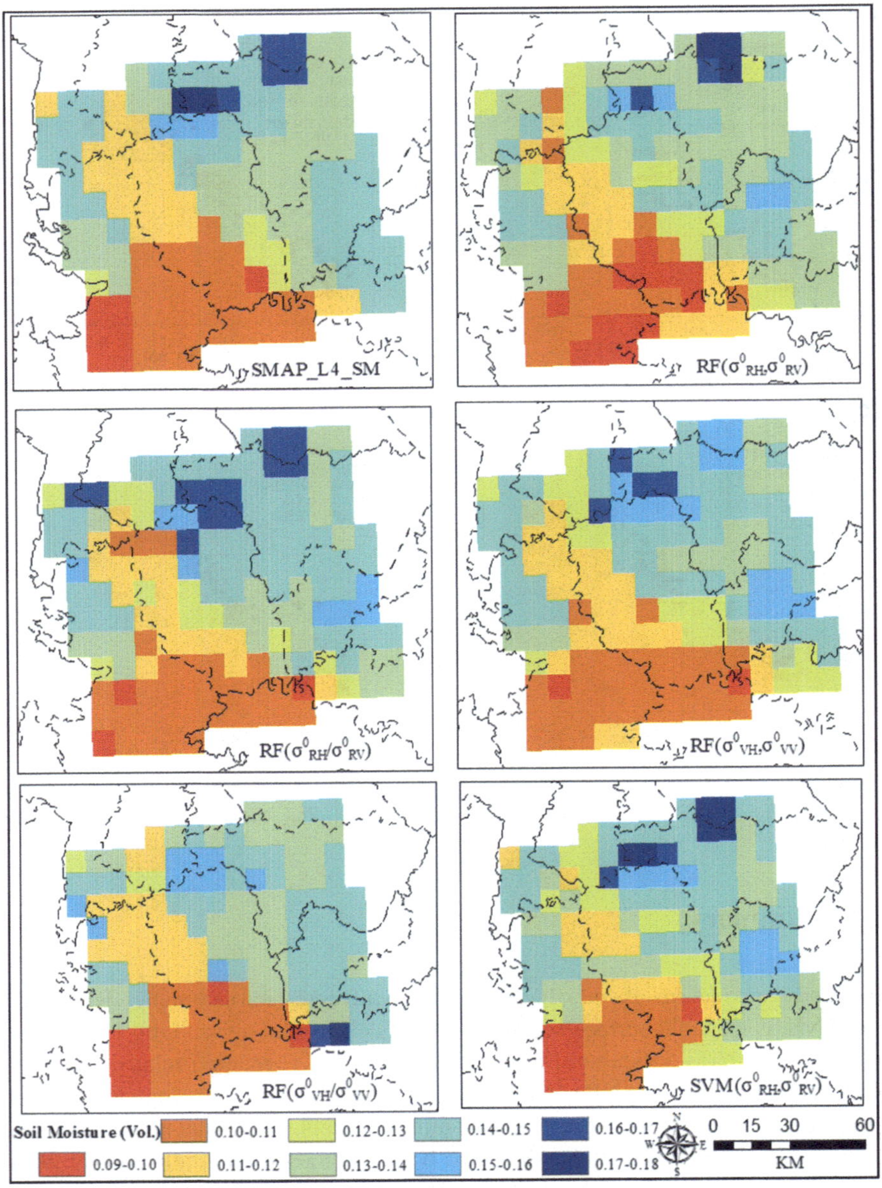

Fig. 6.8 Surface soil moisture map at 9 km grid of **a** SMAP L4_SM; and model predicted using **b** RF with σ^0_{RH}, σ^0_{RV}; **c** RF with $\sigma^0_{RH}/\sigma^0_{RV}$; **d** RF with σ^0_{VH}, σ^0_{VV}; **e** RF with $\sigma^0_{VH}/\sigma^0_{VV}$; **f** SVM with σ^0_{RH}, σ^0_{RV}

σ^0_{RV}) model performed best in estimating the soil moisture content in the region. Other RF models, RF($\sigma^0_{RH}/\sigma^0_{RV}$), RF($\sigma^0_{VH}$, σ^0_{VV}), and RF($\sigma^0_{VH}/\sigma^0_{VV}$), have also performed satisfactorily. In SVM models, SVM(σ^0_{RH}, σ^0_{RV}) performed satisfactorily, However, all other SVM models' performance is not much impressive. The noticeable results for ANN models are only for ANN(σ^0_{RH}, σ^0_{RV}), however, all other ANN models show a poor estimation of soil moisture content.

Similarly, overall, all RF models have shown a higher probability of non-exceedance for less prediction error in comparison to SVM and ANN models. Based on the past research addressing the difficulties associated with measuring soil moisture in a comprehensive manner, high probability of non-exceedance for low error value shows the robustness of the RF models in capturing the variability in soil moisture for the region.

Overall, the models seem to perform comparatively better at higher soil moisture values but do not show reasonable estimates at moderate and low soil moisture values. Most of the scattered samples are of moderate moisture value, therefore have a major impact on model performance. One reason for the unpredictability of moderate soil moisture value may be due to the mature crop cover at that region in the month, which dominates the backscatter signal. A major crop in this region is *mustard* and which is in maturity stage in March month.

In 9 km grid, only RF models show a good correlation with SMAP L4_SM data. Therefore it indicates good prediction stability of RF method in comparison to SVM and good capability of capturing the variability in soil moisture.

It is also observed that within each machine learning method, the model trained using σ^0_{RH}, σ^0_{RV} scored higher value in comparison to others. PONE plot also showed a high probability of non-exceedance for low error value in this combination. All other variables are the same during models development, therefore, it is expected that performance is only governed by backscattered measures. The higher value of performance measures for models using σ^0_{RH}, σ^0_{RV} with other variables emphasized the importance of circularly polarized data in soil moisture prediction. Furthermore, models trained using backscattered bands scored higher value in comparison to their ratio. This result indicated that backscattered bands are more suitable than their ratios.

References

Ahmad, S., Kalra, A., & Stephen, H. (2010). Estimating soil moisture using remote sensing data: A machine learning approach. *Advances in Water Resources, 33*(1), 69–80.

Altese, E., Bolognani, O., & Mancini, M. (1996). Retrieving soil moisture over bare soil from ERS 1 synthetic aperture radar data: Sensitivity analysis based on a theoretical surface scattering model and field data. *Water Resources Research, 32*(3), 653–661.

Altmann, A., Toloşi, L., Sander, O., & Lengauer, T. (2010). Permutation importance: A corrected feature importance measure. *Bioinformatics, 26*(10), 1340–1347.

Álvarez-Mozos, J., González-Audícana, M., & Casalí, J. (2007). Evaluation of empirical and semi-empirical backscattering models for surface soil moisture estimation. *Canadian Journal of Remote Sensing, 33*(3), 176–188.

ASCE Task Committee. (2000a). Artificial neural networks in hydrology. I: Preliminary concepts. *Journal of Hydrologic Engineering, 5*(2), 115–123.
ASCE Task Committee. (2000b). Artificial neural networks in hydrology. II: Hydrologic applications. *Journal of Hydrologic Engineering, 5*(2), 124–137.
Baghdadi, N., Cresson, R., El Hajj, M., Ludwig, R., & La Jeunesse, I. (2012). Estimation of soil parameters over bare agriculture areas from C-band polarimetric SAR data using neural networks. *Hydrology and Earth System Sciences, 16*(6), 1607–1621.
Boerner, R. E. J., Morris, S. J., Sutherland, E. K., & Hutchinson, T. F. (2000). Spatial variability in soil nitrogen dynamics after prescribed burning in Ohio mixed-oak forests. *Landscape Ecology, 15*(5), 425–439.
Breiman, L., Friedman, J. H., Olshen, R. A., & Stone, C. J. (1984). Classification and regression trees. Cole Publishing, Monterey. California, USA: Wadsworth and Brooks/Cole.
Breiman, L. (2001). Randomforest. *Machine Learning, 45*(1), 5–32.
Buyantuyev, A., Wu, J., & Gries, C. (2007). Estimating vegetation cover in an urban environment based on Landsat ETM+ imagery: A case study in Phoenix, USA. *International Journal of Remote Sensing, 28*(2), 269–291.
Cable, J., Kovacs, J., Shang, J., & Jiao, X. (2014). Multi-temporal polarimetric RADARSAT-2 for land cover monitoring in Northeastern Ontario, Canada. *Remote Sensing, 6*(3), 2372–2392.
Chai, S.-S., Walker, J. P., Makarynskyy, O., Kuhn, M., Veenendaal, B., & West, G. (2009). Use of soil moisture variability in artificial neural network retrieval of soil moisture. *Remote Sensing, 2*(1), 166–190.
Chang, C.-C., & Lin, C.-J. (2011). LIBSVM : A library for support vector machines.
Chen, Y., & Gillieson, D. (2009). Evaluation of Landsat TM vegetation indices for estimating vegetation cover on semi-arid rangelands: A case study from Australia. *Canadian Journal of Remote Sensing, 35*(5), 435–446.
Coopersmith, E. J., Cosh, M. H., Bell, J. E., & Boyles, R. (2016). Using machine learning to produce near surface soil moisture estimates from deeper in situ records at U.S. climate reference network (USCRN) locations: Analysis and applications to AMSR-E satellite validation. *Advances in Water Resources, 98*, 122–131.
Cui, Y., Long, D., Hong, Y., Zeng, C., Zhou, J., Han, Z., et al. (2016). Validation and reconstruction of FY-3B/MWRI soil moisture using an artificial neural network based on reconstructed MODIS optical products over the Tibetan Plateau. *Journal of Hydrology, 543*, 242–254.
Das, K., & Paul, P. K. (2015). Present status of soil moisture estimation by microwave remote sensing. *Cogent Geoscience, 1*(1), 1–21.
Dawson, M. S., Fung, A. K., & Manry, M. T. (1997). A robust statistical-based estimator for soil moisture retrieval from radar measurements. *IEEE Transactions on Geoscience and Remote Sensing, 35*(1), 57–67.
Deng, J., Chen, X., Du, Z., & Zhang, Y. (2011). Soil water simulation and predication using stochastic models based on LS-SVM for red soil region of China. *Water Resources Management, 25*(11), 2823–2836.
Desalegn, T., Cruz, F., Kindu, M., Turrión, M. B., & Gonzalo, J. (2014). Land-use/land-cover (LULC) change and socioeconomic conditions of local community in the central highlands of Ethiopia. *International Journal of Sustainable Development & World Ecology, 21*(5), 406–413.
Dobson, M. C., & Ulaby, F. T. (1981). Microwave backscatter dependence on surface roughness, soil moisture and soil texture: Part III: Soil tension. *IEEE Transactions on Geoscience and Remote Sensing, 19*(1), 51–61.
Du, Y., Ulaby, F. T., & Dobson, M. C. (2000). Sensitivity to soil moisture by active and passive microwave sensors. *IEEE Transactions on Geoscience and Remote Sensing, 38*(1), 105–114.
Feng, D., Zhao, Y., Yu, L., Li, C., Wang, J., Clinton, N., et al. (2016). Circa 2014 African land-cover maps compatible with FROM-GLC and GLC2000 classification schemes based on multi-seasonal Landsat data. *International Journal of Remote Sensing, 37*(19), 4648–4664.

Gao, X., Lu, T., Liu, P., & Lu, Q. (2014). A soil moisture classification model based on SVM used in agricultural WSN. *2014 IEEE 7th Joint International Information Technology and Artificial Intelligence Conference* (pp. 432–436). IEEE: Chongqing, China.

Genis, A., Vulfson, L., Blumberg, D. G., Sprinstin, M., Kotlyar, A., Freilikher, V., et al. (2013). Retrieving parameters of bare soil surface roughness and soil water content under arid environment from ERS-1, -2 SAR data. *International Journal of Remote Sensing, 34*(17), 6202–6215.

Gessler, P. E., Moore, I. D., Mckenzie, N. J., & Ryan, P. J. (1995). Soil-landscape modelling and spatial prediction of soil attributes. *International Journal of Geographical Information Systems, 9*(4), 421–432.

Gill, T., Collett, L., Armston, J., Eustace, A., Danaher, T., Scarth, P., et al. (2010). Geometric correction and accuracy assessment of Landsat-7 ETM+ and Landsat-5 TM imagery used for vegetation cover monitoring in Queensland, Australia from 1988 to 2007. *Journal of Spatial Science, 55*(2), 273–287.

Gill, T., Johansen, K., Phinn, S., Trevithick, R., Scarth, P., & Armston, J. (2017). A method for mapping Australian woody vegetation cover by linking continental-scale field data and long-term Landsat time series. *International Journal of Remote Sensing, 38*(3), 679–705.

Giraldo, M. A., Bosch, D., Madden, M., Usery, L., & Kvien, C. (2008). Landscape complexity and soil moisture variation in south Georgia, USA, for remote sensing applications. *Journal of Hydrology, 357*(3–4), 405–420.

Gorthi, S. (2011). *Prediction models for estimation of moisture content*. Logan, Utah: Utah State University.

Hsieh, C.-Y. (2001). Microwave backscattering model for a bare soil field. *Electromagnetics, 21*(3), 259–273.

Hulley, G. C., Hook, S. J., & Baldridge, A. M. (2010). Investigating the effects of soil moisture on thermal infrared land surface temperature and emissivity using satellite retrievals and laboratory measurements. *Remote Sensing of Environment, 114*(7), 1480–1493.

Hutchinson, T. F., Boerner, R. E. J., Iverson, L. R., Sutherland, S., & Sutherland, E. K. (1999). Landscape patterns of understory composition and richness across a moisture and nitrogen mineralization gradient in Ohio (U.S.A.) Quercus forests. *Plant Ecology, 144*(2), 177–189.

Iverson, L. R., Dale, M. E., Scott, C. T., & Prasad, A. (1997). A GIS-derived integrated moisture index to predict forest composition and productivity of Ohio forests (U.S.A.). *Landscape Ecology, 12*(5), 331–348.

Jiang, H., & Cotton, W. R. (2004). Soil moisture estimation using an artificial neural network: A feasibility study. *Canadian Journal of Remote Sensing, 30*(5), 827–839.

Khanna, S., Palacios-Orueta, A., Whiting, M. L., Ustin, S. L., Riaño, D., & Litago, J. (2007). Development of angle indexes for soil moisture estimation, dry matter detection and land-cover discrimination. *Remote Sensing of Environment, 109*(2), 154–165.

Laamanen, R., & Verlinden, A. (2006). Modeling woody vegetation resources using Landsat TM imagery in northern Namibia. *The Southern African Forestry Journal, 207*(1), 27–39.

Lakhankar, T., Ghedira, H., Temimi, M., Sengupta, M., Khanbilvardi, R., & Blake, R. (2009). Non-parametric methods for soil moisture retrieval from satellite remote sensing data. *Remote Sensing, 1*(1), 3–21.

Lakshmi, V. (2013). Remote sensing of soil moisture. *ISRN Soil Science*, 1–33.

Li, E., Du, P., Samat, A., Xia, J., & Che, M. (2015). An automatic approach for urban land-cover classification from Landsat-8 OLI data. *International Journal of Remote Sensing, 36*(24), 5983–6007.

Li, J., & Islam, S. (1999). On the estimation of soil moisture profile and surface fluxes partitioning from sequential assimilation of surface layer soil moisture. *Journal of Hydrology, 220*(1–2), 86–103.

Liaw, A., & Wiener, M. (2002). *Classification and Regression by Random Forest, 2*(3), 18–22.

Liu, D., Yu, Z., & Lü, H. (2010). Data assimilation using support vector machines and ensemble Kalman filter for multi-layer soil moisture prediction. *Water Science and Engineering, 3*(4), 361–377.

Liu, H., Xie, D., & Wu, W. (2008). Soil water content forecasting by ANN and SVM hybrid architecture. *Environmental Monitoring and Assessment, 143*(1–3), 187–193.

Liu, Y., Chen, K. S., Xu, P., & Li, Z. L. (2016a). Modeling and characteristics of microwave backscattering from rice canopy over growth stages. *IEEE Transactions on Geoscience and Remote Sensing, 54*(11), 6757–6770.

Liu, Y., Zeng, J., Chen, K.-S., & Li, Z. (2016b). Parameter sensitivity analysis for bistatic scattering of rough surface. *2016 IEEE International Geoscience and Remote Sensing Symposium (IGARSS)* (pp. 4335–4338). IEEE: Beijing, China.

Loew, A. (2008). Impact of surface heterogeneity on surface soil moisture retrievals from passive microwave data at the regional scale: The Upper Danube case. *Remote Sensing of Environment, 112*(1), 231–248.

Louppe, G., Wehenkel, L., Sutera, A., & Geurts, P. (2013). Understanding variable importances in forests of randomized trees. *Advances in Neural Information Processing Systems, 26*, 431–439.

Mallick, K., Bhattacharya, B. K., & Patel, N. K. (2009). Estimating volumetric surface moisture content for cropped soils using a soil wetness index based on surface temperature and NDVI. *Agricultural and Forest Meteorology, 149*(8), 1327–1342.

McBratney, A. B., Mendonça Santos, M. L., & Minasny, B. (2003). On digital soil mapping. *Geoderma, 117*(1–2), 3–52.

MirMazloumi, S. M., & Sahebi, M. R. (2016). Assessment of different backscattering models for bare soil surface parameters estimation from SAR data in band C, L and P. *European Journal of Remote Sensing, 49*(1), 261–278.

Mishra, M. D., Patel, P., Srivastava, H. S., Patel, P. R., Shukla, A., & Shukla, A. K. (2014). Absolute radiometric calibration of FRS-1 and MRS mode of RISAT-1 synthetic aperture radar (SAR) data using corner reflectors. *International Journal of Advanced Engineering Research and Science, 1*(6), 78–89.

Morris, S. J., & Boerner, R. E. J. (1998). Landscape patterns of nitrogen mineralization and nitrification in southern Ohio hardwood forests. *Landscape Ecology, 13*(4), 215–224.

Munyati, C., & Sinthumule, N. I. (2013). Assessing change in woody vegetation cover in the Kruger National Park, South Africa, using spectral mixture analysis of a Landsat TM image time series. *International Journal of Environmental Studies, 70*(1), 94–110.

Muramatsu, K., Xiong, Y., Nakayama, S., Ochiai, F., Daigo, M., Hirata, M., et al. (2007). A new vegetation index derived from the pattern decomposition method applied to Landsat-7/ETM+ images in Mongolia. *International Journal of Remote Sensing, 28*(16), 3493–3511.

Myers, W., Linden, S., & Wiener, G. (2009). A data mining approach to soil temperature and moisture prediction. *7th Conference on Artificial Intelligence and Its Applications to the Environmental Sciences* (pp. 1–4). Phoenix: Arizona.

Oppel, S., Strobl, C., & Huettmann, F. (2009). *Alternative methods to quantify variable importance in ecology*. Munich.

Palacios-Orueta, A., Khanna, S., & Litago, J. (2005). Assessment of NDVI and NDWI spectral indices using MODIS time series analysis and development of a new spectral index based on MODIS shortwave infrared bands. In *1st International Conference of Remote Sensing and Geoinformation Processing*. Trier, Germany.

Paloscia, S., Pampaloni, P., Pettinato, S., & Santi, E. (2008). A comparison of algorithms for retrieving soil moisture from ENVIS AT/AS AR images. *IEEE Transactions on Geoscience and Remote Sensing, 46*(10), 3274–3284.

Paloscia, S., Pettinato, S., Santi, E., Notarnicola, C., Pasolli, L., & Reppucci, A. (2013). Soil moisture mapping using Sentinel-1 images: Algorithm and preliminary validation. *Remote Sensing of Environment, 134*, 234–248.

Pasolli, L., Notarnicola, C., Bruzzone, L., Bertoldi, G., Chiesa, S. D., Niedrist, G., et al. (2012). Polarimetric RADARSAT-2 imagery for soil moisture retrieval in alpine areas. *Canadian Journal of Remote Sensing, 37*(5), 535–547.

Patel, P., Srivastava, H. S., Panigrahy, S., & Parihar, J. S. (2006). Comparative evaluation of the sensitivity of multi-polarized multi-frequency SAR backscatter to plant density. *International Journal of Remote Sensing, 27*(2), 293–305.

Purevdorj, T., Tateishi, R., Ishiyama, T., & Honda, Y. (1998). Relationships between percent vegetation cover and vegetation indices. *International Journal of Remote Sensing, 19*(18), 3519–3535.

Ramírez-villegas, J., & Bueno-cabrera, A. (2009). *Working with climate data and niche modeling: Creation of bioclimatic variables*. Colombia: Cali.

Rodriguez-galiano, V. F., Ghimire, B., Rogan, J., Chica-olmo, M., & Rigol-sanchez, J. P. (2012). An assessment of the effectiveness of a random forest classifier for land-cover classification. *ISPRS Journal of Photogrammetry and Remote Sensing, 67*, 93–104.

Said, S., Kothyari, U. C., & Arora, M. K. (2008). ANN-based soil moisture retrieval over bare and vegetated areas using ERS-2 SAR data. *Journal of Hydrologic Engineering, 13*(6), 461–475.

Satalino, G., Mattia, F., Davidson, M. W. J., Le Toan, T., Pasquariello, G., & Borgeaud, M. (2002). On current limits of soil moisture retrieval from ERS-SAR data. *IEEE Transactions on Geoscience and Remote Sensing, 40*(11), 2438–2447.

Shi, J., Wang, J., Hsu, A. Y., O'Neill, P. E., & Engman, E. T. (1997). Estimation of bare surface soil moisture and surface roughness parameter using L-band SAR image data. *IEEE Transactions on Geoscience and Remote Sensing, 35*(5), 1254–1266.

Shukla, G., Garg, R. D., Srivastava, H. S., & Garg, P. K. (2017). Implementation of random forest algorithm for crop mapping across an aridic to ustic area of Indian states. *Journal of Applied Remote Sensing, 11*(2), 26005.

Shukla, G., Garg, R. D., Srivastava, H. S., & Garg, P. K. (2018). Performance analysis of different predictive models for crop classification across an aridic to ustic area of Indian states. *Geocarto International, 33*(3), 240–259.

Song, X., Zhang, G., Liu, F., Li, D., Zhao, Y., & Yang, J. (2016). Modeling spatio-temporal distribution of soil moisture by deep learning-based cellular automata model. *Journal of Arid Land, 8*(5), 734–748.

Srivastava, P. K., Han, D., Ramirez, M. R., & Islam, T. (2013). Machine learning techniques for downscaling SMOS satellite soil moisture using MODIS land surface temperature for hydrological application. *Water Resources Management, 27*(8), 3127–3144.

Strobl, C., Boulesteix, A.-L., Zeileis, A., & Hothorn, T. (2007). Bias in random forest variable importance measures: Illustrations, sources and a solution. *BMC Bioinformatics, 8*(1), 25.

Su, Z., Troch, P. A., & DeTroch, F. P. (1997). Remote sensing of bare surface soil moisture using EMAC/ESAR data. *International Journal of Remote Sensing, 18*(10), 2105–2124.

Townshend, J.R., Masek, J.G., Huang, C., Vermote, E.F., Gao, F., Channan, S., et al. (2012). Global characterization and monitoring of forest cover using Landsat data: Opportunities and challenges. *International Journal of Digital Earth, 5*(5), 373–397.

Ulaby, F. T., Bradley, G. A., & Obson, M. C. (1979). Microwave backscatter dependence on surface roughness, soil moisture, and soil texture: Part II-vegetation covered soil. *IEEE Transactions on Geoscience Electronics, 17*(2), 33–40.

Ulaby, F. T., Razani, M., & Dobson, M. C. (1983). Effects of vegetation cover on the microwave radiometric sensitivity to soil moisture. *IEEE Transactions on Geoscience and Remote Sensing, GE-21*(1), 51–61.

Vapnik, V. N. (2000). The nature of statistical learning theory. In M. Jordan, S. L. Lauritzen, J. F. Lawless, & V. Nair (Eds.), *Statistics for engineering and information science* (pp. 1564–1564). Springer: Verlag, New York.

Verhoest, N. E. C., Lievens, H., Wagner, W., Álvarez-Mozos, J., Moran, M. S., & Mattia, F. (2008). On the soil roughness parameterization problem in soil moisture retrieval of bare surfaces from synthetic aperture radar. *Sensors, 8*(7), 4213–4248.

Wagner, W., & Scipal, K. (2000). Large-scale soil moisture mapping in western Africa using the ERS scatterometer. *IEEE Transactions on Geoscience and Remote Sensing, 38*(4), 1777–1782.

Waldteufel, P., Richaume, P., Kerr, Y., Wigneron, J. P., Mahmoodi, A., Mialon, A., et al. (2007). Optimizing the algorithm for retrieving soil moisture from SMOS data. In *IEEE International Geoscience and Remote Sensing Symposium, 2007. IGARSS 2007* (pp. 3952–3954). Barcelona, Spain.

Wang, L., & Qu, J. J. (2009). Satellite remote sensing applications for surface soil moisture monitoring: A review. *Frontiers of Earth Science in China, 3*(2), 237–247.

Wigneron, J. P., Calvet, J. C., Pellarin, T., Van De Griend, A. A., Berger, M., & Ferrazzoli, P. (2003). Retrieving near-surface soil moisture from microwave radiometric observations: Current status and future plans. *Remote Sensing of Environment, 85*(4), 489–506.

Witten, I. H., & Frank, E. (2005). Data mining: Practical machine learning tools and techniques (Google eBook) (2nd ed.). Complementary literature None. San Francisco: Elsevier.

Xie, X. M., Xu, J. W., Zhao, J. F., Liu, S., & Wang, P. (2014). Soil moisture inversion using AMSR-E remote sensing data: An artificial neural network approach. *Applied Mechanics and Materials, 501–504*, 2073–2076.

Xin, Q., Broich, M., Zhu, P., & Gong, P. (2015). Modeling grassland spring onset across the Western United States using climate variables and MODIS-derived phenology metrics. *Remote Sensing of Environment, 161*, 63–77.

Yang, S., & Huang, Y. (2002). Application of support vector machine based on time series for soil moisture and nitrate nitrogen content prediction. In D. Li & Z. Chunjiang (Eds.), *IFIP Advances in Information and Communication Technology* (pp. 2037–2045). Boston: Springer.

Zaman, B., McKee, M., & Neale, C. M. U. (2012). Fusion of remotely sensed data for soil moisture estimation using relevance vector and support vector machines. *International Journal of Remote Sensing, 33*(20), 6516–6552.

Zhang, D., Zhang, W., Huang, W., Hong, Z., & Meng, L. (2017). Upscaling of surface soil moisture using a deep learning model with VIIRS RDR. *ISPRS International Journal of Geo-Information, 6*(5), 130.

Zhao, S., Yang, Y., Qiu, G., Qin, Q., Yao, Y., Xiong, Y., et al. (2010). Remote detection of bare soil moisture using a surface-temperature-based soil evaporation transfer coefficient. *International Journal of Applied Earth Observation and Geoinformation, 12*(5), 351–358.

The manufacturer's authorised representative in the EU is Springer Nature Customer Service Centre GmbH, Europaplatz 3, 69115 Heidelberg, Germany. If you have any concerns regarding our products, please contact ProductSafety@springernature.com

Printed and bound by CPI Group (UK) Ltd, Croydon, CR0 4YY
25/09/2026
02222901-0020